8th Edition
National Express
Coach Handbook

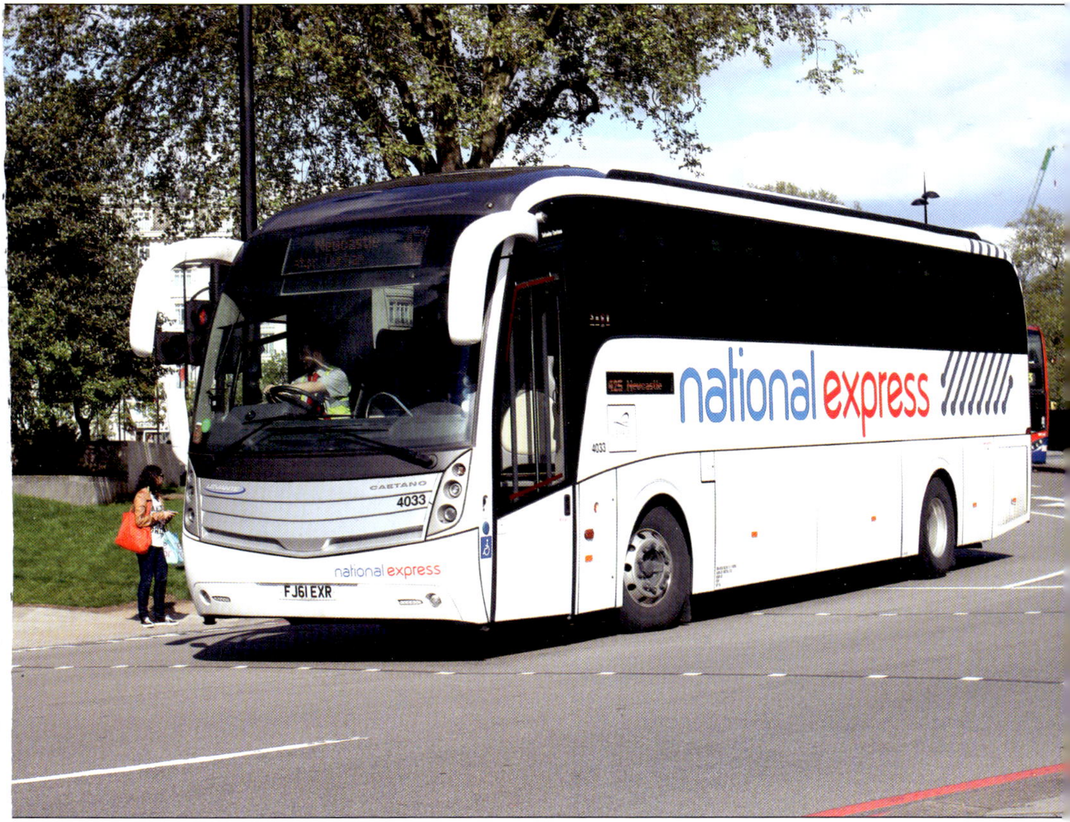

British Bus Publishing

Body codes used in the Bus Handbook series:
Type:
A	Articulated vehicle
B	Bus, either single-deck or double-deck
BC	Interurban - high-back seated bus
C	Coach
M	Minibus with design capacity of 16 seats or less
N	Low-floor bus (Niederflur), either single-deck or double-deck
O	Open-top bus (CO = convertible - PO = partial open-top)

Seating capacity is then shown. For double-decks the upper deck quantity is followed by the lower deck.

Please note that seating capacities shown are generally those provided by the operator. It is common practice, however, for some vehicles to operate at different capacities when on certain duties.

Door position:-
C	Centre entrance/exit
D	Dual doorway.
F	Front entrance/exit
R	Rear entrance/exit (no distinction between doored and open)
T	Three or more access points

Equipment:-
T	Toilet	TV	Training vehicle.
M	Mail compartment	RV	Used as tow bus or engineers' vehicle.

Allocation:-
s	Ancillary vehicle
t	Training bus
u	out of service or strategic reserve; refurbishment or seasonal requirement
w	Vehicle is withdrawn and awaiting disposal.

e.g. - B32/28F is a double-deck bus with thirty-two seats upstairs, twenty-eight down and a front entrance/exit., N43D is a low-floor bus with two or more doorways.

Re-registrations:-

Where a vehicle has gained new index marks the details are listed at the end of each fleet showing the current mark, followed in sequence by those previously carried starting with the original mark.

Annual books are produced for the major groups:
The Stagecoach Bus Handbook
The First Bus Handbook
The Arriva Bus Handbook
The Go-Ahead Bus Handbook
The National Express Coach Handbook
Some editions for earlier years are available. Please contact the publisher.

Regional books in the series:
The Scottish Bus Handbook
The Welsh Bus Handbook
The Ireland & Islands Bus Handbook
English Bus Handbook: Smaller Groups
English Bus Handbook: Notable Independents
English Bus Handbook: Coaches

Associated series:
The Hong Kong Bus Handbook
The Malta Bus Handbook
The Leyland Lynx Handbook
The Postbus Handbook
The Mailvan Handbook
The Toy & Model Bus Handbook - Volume 1 - Early Diecasts
The Fire Brigade Handbook (fleet list of each local authority fire brigade)
The Police Range Rover Handbook

Some earlier editions of these books are still available. Please contact the publisher on 01952 255669.

Contents

History	5	Mike de Courcey	35
		Park's of Hamilton	36
National Express	16	Peter Godward Coaches	38
Ace Travel	20	Premiere	38
Ambassador Travel	20	Rotala	40
Bayliss Ecexutive	20	Selwyns	41
Bennett's	21	Silverdale	43
BL Travel	21	Skyline Travel	44
Bruce's	22	South Gloucestershire	44
Burgin European	22	Stagecoach	46
Buzzlines Travel	22	E Stott and Sons	47
Chalfont	23	Stuart's of Carluke	48
R W Chenery	24	Tellings-Golden Miller	49
East Yorkshire	25	Travelstar	50
Edwards Coaches	26	Travellers' Choice	51
Elcock Reisen	27	Whittles	52
Epsom Coaches	27	Yellow Buses	53
Excelsior	28	Yeomans	54
Galloway	29	Yourbus	55
Go North East	30		
Go South Coast	31	National Express Services	57
Johnson Bros	32		
The King's Ferry	32	Vehicle Index	59
Llew Jones	33		
Lucketts	34		
Mainline	34		

The National Express Coach Handbook

The National Express Coach Handbook is part of the Bus Handbook series that details the fleets of selected bus and coach operators. These Bus Handbooks are published by British Bus Publishing. Although this book has been produced with the encouragement of, and in co-operation with National Express management, it is not an official publication. The vehicles included are subject to variation, particularly as new vehicle deliveries lead to older vehicles being withdrawn. The contents are correct to August 2011.

The Bus Handbook series is concerned primarily with vehicles this volume features coaches operated on National Express services. The National Express bus operations are included in the 'English Bus Handbook - Groups' book.

Quality photographs for inclusion in the series are welcome, for which a fee is paid. Unfortunately the publishers cannot accept responsibility for any loss and they require that you show your name on each picture or slide. High-resolution digital images of six megapixels or higher are also welcome on CD or DVDs.

To keep the fleet information up to date we recommend the magazine, Buses, published monthly by Key Publications, or for more detailed information, the PSV Circle monthly news sheets. The writer and publisher would be glad to hear from readers should any information be available which corrects or enhances that given in this publication.

Principal Editor: Stuart Martin.

Acknowledgments: We are grateful to Brian Bannister, Stephen Byrne, Keith Grimes, Tom Johnson, Mike Lambden, Colin Lloyd, Mark Lyons, Malcolm Tranter, the PSV Circle and the management and officials of National Express and EuroLines for their kind assistance and co-operation in the compilation of this book.

The front cover photograph and frontispiece are by Dave Heath while the rear cover views are by Mark Doggett.

1st Edition	2000	ISBN 1-897990-56-1
2nd Edition	2002	ISBN 1-897990-58-8
3rd Edition	2004	ISBN 1-904875-04-1
4th Edition	2006	ISBN 1-904875-49-1
5th Edition	2008	ISBN 9781904875505
6th Edition	2009	ISBN 9781904875482
7th Edition	2011	ISBN 9781904875475

ISBN 9781904875 78 9   © Published by British Bus Publishing Ltd, September 2012

British Bus Publishing Ltd, 16 St Margaret's Drive, Telford, TF1 3PH

Telephone: 01952 255669

web; www.britishbuspublishing.co.uk
e-mail: sales@britishbuspublishing.co.uk

NATIONAL EXPRESS

National Express Ltd, National Express House,
Birmingham Coach Station, Mill Lane, Birmingham B5 6DD

A full history of National Express

Although stagecoaches were undoubtedly the forerunners of the present long-distance coach network it was not until after the First World War and the introduction of motor buses that express coach services really came into their own.

In 1919, Elliott Bros., whose coaches carried the famous 'Royal Blue' fleet name, and who had run one of the earlier horse-drawn services, introduced a limited form of express coach service operating between Bournemouth and London. However, Greyhound Motors of Bristol is generally acknowledged as being the first to introduce a daily, all year round, motorised express coach service in Britain. This service which was launched in 1925, linked Bristol with London and expanded rapidly. Many other operators, able to see the commercial benefits of long-distance travel, began similar services in the following months.

The 1930 Road Traffic Act introduced a system of licensing that covered drivers, conductors and the routes that were operated and successfully brought order to a chaotic, rapidly growing, and somewhat haphazard industry. Intending bus and coach operators now found it much harder to introduce new services, with each application for a new or revised service requiring a lengthy process to the local government appointed Traffic Commissioner. This new system of licensing provided the stability for expansion and early co-operation amongst coach operators gave rise to the formation of the first networks of co-ordinated services.

These 'Pool' networks greatly increased travel opportunities for the rapidly growing number of coach passengers. On 1 July 1934, Elliott Brothers became a founder member of the Associated Motorways pool based at Cheltenham. The company's services to the Midlands became AM services and were operated by Royal Blue coaches. Another 'pool' was London Coastal Coaches, based at Victoria Coach Station, which had opened two years earlier in 1932, replacing the original 'London' terminus in Lupus Street of 1924 vintage.

Although most express coaches were suspended in 1942 for the remaining years of the Second World War, some Royal Blue services continued to run to help serve southern areas that were not well provided for by local bus services. After the war, the full network gradually resumed from April 1946 with new coach designs helping to increase passenger numbers.

A new era dawns

The steady increase in the coach passengers peaked in the late 1950s followed by a gradual decline due to the increase in the number of private cars. In 1959 the opening of the first stretches of Britain's new motorway network brought new opportunities for coach operators such as Midland Red of Birmingham and Ribble of Preston which introduced the Gay Hostess double-deck coaches.

By the late 1960s most bus companies, with the exception of municipal and small independent operators had formed into two main groups, the state-owned Tilling Group and the British Electric Traction Group (BET). In March 1968, the government brought both groups together under the Transport Holding Company.

The 1968 Transport Act brought about an integrated public passenger transport system across the country. One of the major provisions of the Act was the formation, on 28 November 1968, of the National Bus Company (NBC). NBC began operating on 1 January 1969 and, by 31 December 1969, NBC controlled ninety-three bus companies grouped into forty-four operating units employing 81,000 staff and having a fleet of 21,000 vehicles. A new era of public transport had arrived.

Network developments

From the beginning, the directors of what was the biggest road passenger transport operation in Europe began to bring together the coaching activities of each constituent operator. The reasons were obvious. Each local company was pursuing its own policy of express coach service operation.

Inevitably this was leading to duplication of services and it was soon decided that a co-ordinated policy of express coach service planning would be of benefit to both the customer and the National Bus Company alike. However, regulation of services prevented any real expansion of services or the provision of routes where there were mass markets.

The 'National' brand name was introduced during 1972 and the original 'all white' livery began to appear on coaches as a first stage in offering customers a nationwide standard and a recognisable product. The winter of 1973-74 saw the publication of the first comprehensive coach timetable that included details of the entire 'National' network.

The brand name, National Express, first appeared on publicity in 1974 and on vehicles in 1978. Oddly, some services, such as those from Oxford to London, were not included in the National network.

In 1979, NBC commissioned a major programme of market research called 'Coachmap'. Every passenger on every journey was asked where, when and why he or she was travelling. The substantial amount of information obtained gave a much-needed insight into the travel requirements of both young and old, but was never actually implemented, as the 1980 Transport Act altered the whole of the network.

Deregulation and expansion

The introduction of the 1980 Transport Act on 6th October swept away fifty years of licensing restrictions and introduced competition on long-distance coach routes.

National Express, and the main Scottish express coach operator, Scottish Citylink, faced new competition from a host of established bus and coach operators trying their hand at operating regular long-distance coach services. It came as no surprise to National Express to discover that many of the 'new' operators seemed to want to run coaches only at the busiest times and only on the most popular routes. The future of the nationwide coach network, and of National Express itself, was in jeopardy.

Totally without subsidy, and by introducing new services and lower fares, National Express fought to prevail or perish in the ensuing war. Most of the new operators were unable to sustain viable operation and withdrew from operating their services within a matter of months. Even the co-operative venture mounted countrywide under the title 'British Coachways' failed to capture sufficient business.

The strengths of the nationwide, co-ordinated network operated by National Express became all too apparent and the publicity surrounding the coach war gave a major boost to the long-term fortunes of National Express. Passengers also benefited from the new services and lower fares and the skirmish gave National Express valuable experience that was to prove useful in the years to come. Most importantly National Express was free to provide coach services wherever it felt that there was a market.

Annual passenger figures for the nationwide express coach network increased from 8.5 million in 1979 to around fifteen million in 1986 as a direct result of post-deregulation competition.

Today, the annual figure is around twenty one million and has continued to grow over the last few years. The main difference now is that the summer peaks that were experienced on some of the more popular services have now disappeared and all the UK coach services are now much busier throughout the whole of the year.

'Rapide' growth

With skilful marketing and an eye for the needs of the customer, a handful of independent coach operators fared better than most. Both Trathens from the West Country and Cotters from Scotland (later to become Stagecoach) introduced up-market services operated by coaches carrying hostesses, refreshments and toilets.

Seeing the opportunities that such an operation would present on other services, National Express entered into an agreement with Trathens to co-operate in running the West Country services. This new concept of improved customer care and service quality was given the name 'Rapide'. The Rapide service introduced a hostess/steward service of light refreshments to each seat. The coaches used on the service were fitted with their own toilet/washroom, air suspension and reclining seats. The on-board facilities cut

Cambridge is the location for this view of FJ08KMA working service 787. The vehicle is currently operated by Yeomans and can be found on the Hereford service. *Mark Doggett*

out the need for time consuming refreshment and toilet stops, offering an instant saving in journey times of around twenty percent.

The introduction of Rapide services also brought about the first 'seat reservation' system for National Express. Rapide was launched in 1984 and allowed passengers to reserve seats on specific services. Seat reservations were a revolution in public transport at that time as free-sale bookings often led to overloads. This booking system was further developed in 1987 as a full reservation system across the whole National Express network and renamed 'EXTRA'. It was completely redeveloped in 1998 as part of a Year 2000 project.

Public demand for the new Rapide services was high and brought about the introduction of a new design of double-deck coach to cater for the higher number of customers discovering the many benefits of coach travel for the first time. The demand for on-board catering was seen to be declining in the late 1990s with on-board customer surveys showing that passengers were choosing to bring their own style of refreshments with them for their journey. This, coupled with improvements to catering outlets at key coach and bus stations, resulted in a gradual withdrawal of this facility over a number of years. The on-board catering facility on the few remaining National Express services to offer this service were withdrawn at the start of the 2001 Summer timetable.

The continuing provision of on-board washroom/toilet facilities on all National Express coach services meant that in many cases the running time of these service was unaffected by these changes.

'National Express' is born

On 26 October 1986, following the introduction of the 1985 Transport Act there was deregulation within the industry to all local bus services. Although designed to increase competition between all bus and coach operators there was surprisingly little change in the long-distance express coach market, itself deregulated back in 1980.

However, of greater importance to National Express was the requirement that the National Bus Company should be sold into the private sector. The first subsidiary, National Holidays, was sold in July 1986; the last, London Country (North East) in April 1988.

National Express itself was the subject of a management buy-out, led by Clive Myers, on 17 March 1988. Between 1988 and 1991, National Express Holdings Ltd, the name of the company set up to buy National Express from the National Bus Company, acquired the established North Wales bus and coach operator, Crosville Wales, the Merseyside based coach operator, Amberline, the ATL Holdings Group (which included the Carlton PSV vehicle dealership and the Yelloway Trathen bus and coach company mentioned earlier) and the express coach services of Stagecoach Holdings Ltd based in Perth.

It was during this period that National Express, Plaxton and Volvo created a new purpose-built coach, the Expressliner, which was unveiled on 20 March 1989. The Expressliner, with a kneeling suspension and many other features unique to National Express, brought a new standard of high quality to coach travel across all routes as well as offering the customers a standard product.

This was followed by a second generation Expressliner in 2002 which offered more choice in some of the mechanical features for operators, but still provided the standardisation and comfort for customers.

A Highland 'fling'

The acquisition of the express coach services of the Stagecoach Holdings Group on 31 July 1989, came at the same time as the long-standing agreement with Scottish Citylink coaches on joint operation across the English/Scottish border came to an end. A new National Express network within Scotland was then introduced under a new brand called Caledonian Express. With Head Offices based at the old Stagecoach premises at Walnut Grove in Perth, the Caledonian Express services linked into the main National Express network and the number of new passengers began to grow immediately. New double-deck coaches entered service on the prestigious Rapide services linking London with Scotland and new marketing initiatives were introduced offering a high quality of coach service to and from Scotland for the first time.

In 1993 National Express Group also acquired Scottish Citylink and absorbed the Caledonian Express services into it. This acquisition enabled the Group to offer a truly 'national' coach network with services operating

A recent arrival for the main National Express fleet is FJ12FYR, a Volvo B9R with Caetano Levanté bodywork. It is seen passing through Stratford while operating service A9 to Stansted Airport. The Lavante was designed for, with, and named by, National Express and when launched in 2005 led the luxury coach market with its wheelchair accessibility. *Keith Grimes*

throughout England, Scotland and Wales. However, in August 1998, following the award in April of the franchise to operate ScotRail, Scotland's national railway, National Express Group disposed of Scottish Citylink to Metroline in a deal which left the operation intact and which guaranteed the continuation of cross-border coach travel for a period of time.

The National Express Group 'ScotRail' franchise expired in October 2004 and now National Express operates frequent long distance coaches between the main Scottish cities to all parts of the UK, with connections at London onto the wider European destinations offered by the Eurolines network.

Coaches can 'float'

Throughout its long and varied history, National Express has faced many changes. On 23 July 1991, a consortium made up of a number of City investment companies and the Drawlane Transport Group bought out National Express Holdings Ltd.

The chairman of the Drawlane Transport Group, Ray McEnhill, moved from that position and became the Chief Executive of the new company, the National Express Group Limited. Crosville Wales and Amberline were not included in the deal. On 1 December 1992, National Express took another change of direction when Chief Executive Ray McEnhill and deputy chief executive Adam Mills led National Express Group on to the Stock Market through the London Stock Exchange at a share price of 165p.

The prospectus issued at the time of the flotation made the Group's new strategy for development clear. Its objectives were to re-focus and improve the profitability of the core coach business, develop new products and services within its existing operations and acquire new businesses in the passenger transport market. It was during this period that development of a centralised call-centre structure, with one national number, commenced and the first steps taken on establishing what is now the very successful website.

Group growth

The National Express Group's policy is to expand the group further by acquisitions within associated areas of the travel industry. This expansion continues to take place not only in the UK but also within Europe and overseas. The National Express Group currently has interests in the UK, North America, Spain and Morocco.

United Kingdom

The Group has expansive operations in Coach, Bus and Train operating companies within the UK. These include coach brands such as National Express, Airlinks and Eurolines and bus operations that include West Midlands, Coventry, Dundee, and the Midland Metro. National Express Group also operates c2c, the UK's most reliable rail service. In November 2007, The Kings Ferry company was acquired. It is known for its high-quality coach hire and commuter services from Kent to London and continues to operate with its own identity.

North America

In North America the company also operates Durham School Services which provides student transportation throughout the USA and Stock Transportation which provides student transportation in two provinces of Canada.

Spain

In January 2006 National Express Group also acquired Alsa, a transport company which provides coach and bus services throughout Spain with operations also in Portugal and Morocco. Subsequently Continental Auto was also acquired making National Express by far the largest provider of express coach services in Spain.

Coach stations

During 1994 the first purpose-built coach station to be constructed in Britain for over twenty-five years was opened in Norton Street, Liverpool. This new facility, which was widely acclaimed, greatly increased the number

In addition to the standard scheme, several coaches carry colours for special events. FJ60EFV, operated by Selwyns, is seen passing through Hanley bus station while carrying England's national flag livery commemorating National Express' association with the Wembley Stadium. *Cliff Beeton*

of customers using National Express coach services from Merseyside and its environs.

Similar increases in passenger numbers were to be seen when new coach stations were opened by National Express at Leeds in 1996, Southampton in 1998, Manchester Central in 2002, Newcastle-upon-Tyne during Spring 2003 and Heathrow's impressive new Central Bus Station which opened in Spring 2006. The jewel in the crown, is the new Birmingham Coach Station, which was opened in December 2009 by Fabio Capello, the then England football manager. Whilst not constructed by National Express, the company is also operating the new Milton Keynes Coachway which opened in December 2010.

Ticketing

Improvements have also been made to assist customers who want to find out about National Express services and who then want to make credit card bookings by phone, at kiosks or using the internet. The new Customer Contact Centre, opened by the Minister of Transport John Spellar in July 2001, based in central Birmingham, offers passengers the very latest in Call Centre Technology. Centralising this activity on just one site offered major benefits to passengers and enabled National Express to offer an improved customer service. It has never been easier for customers to book a National Express ticket. Customers can contact the call centre, visit the website, buy tickets from ticket desks or ticket kiosks at stations or visit any one of the

Originally only available on the Volvo B12B chassis, the Levanté soon became available on the Scania K series chassis and more recently on the Volvo B9 and MAN products. Norwich is the location for this view of National Express 68, FH08EBM, an example based on the B12B. *Dave Heath*

800 National Express ticket agents around the UK. It is now a truly 24/7 sales operation as from 2011 the Call Centre went to 24 hour operation.

National Express was also one of the first UK travel companies to recognise the importance of the internet for customers wishing to obtain both travel information and to book coach tickets at any time of the day, and from anywhere in the world. Customers booking on-line also benefit from Funfares which offer travel from just £1 per single journey and are now available on most of the main services. Online sales are the most important sales mode. The nationalexpress.com website now handles thirty-five million page views per month and up to one and a half million viewings a day, and has won many awards over the last few years. The introduction of 'e-tickets' in 2002 was warmly welcomed by National Express Limited (NEL) customers, allowing them to both book and print their own coach ticket from the comfort of their home or work pc with coach drivers simply checking the unique reference number issued with each booking.

During 2006, m-Tickets were also introduced. This innovative ticketing solution allowed customers to have the opportunity of booking their tickets by phone or the website and then to receive their ticket as a text message which is shown to the driver on boarding.

Airports take-off

Airport coach services have always been an important part of the National Express business and in October 1994 a newly branded service called

Airlinks was introduced specifically for the growing airport market. It was first established on the Bradford/Leeds to Heathrow/Gatwick corridor. This was followed in May 1995 with the introduction of Airlinks services on corridors between Newcastle/Nottingham to Heathrow/Gatwick, Swansea/Cardiff to Heathrow/Gatwick and Bristol to Heathrow/Gatwick. Early in 1996 the acquisition of the Flightlink brand saw the inclusion of new airport corridors from the West Midlands to Heathrow, Gatwick and Manchester airports. This was followed by the re-branding of all dedicated airport corridors to Flightlink and the launch of the Flightlink network to the retail travel trade.

In mid-1997 Speedlink Airport Services commenced operation of Hotel Hoppa, serving all thirteen hotels located around Heathrow. This major operation, using thirty low-floor buses was a major partnership between Speedlink, BAA and the Heathrow airport hoteliers, and succeeded in reducing traffic congestion in the Heathrow central area by over 30%.

Following a decision made in mid-1998 to bring together the airport operations of Speedlink Airport Services Ltd and the NEL airport services brand of Flightlink, a new company was formed on 1 January 1999 called AirLinks the Airport Coach Company Limited. Its aim was to focus on airport-scheduled and contract bus and coach services and operated vehicles with many distinctive liveries such as Flightlink, Speedlink and Jetlink. The airport coach service network continued to grow with AirLinks acquiring all third-party interests in the Jetlink brand, Silverwing Transport Services, Cambridge Coach Services Ltd, Airbus and Capital Logistics; all of which provided coach and bus operations within the Stansted, Luton, Heathrow and Gatwick airports. AirLinks soon became the largest operator of both scheduled and contract services to BAA and the airline operators. But changes to these airport services and other National Express routes were soon to take place.

Revised look

After many years of acquiring different coach businesses, National Express had actually become an organisation that was operating under many different brand names. Flightlink, Jetlink, Speedlink, Express Shuttle, GoByCoach, Airbus seemed at times to be competing with one another. Something needed to happen to bring all these businesses together.

On 03/03/03 National Express revealed its then new corporate identity which included a revised logo and coach livery. The launch, which coincided with the thirtieth anniversary of National Express, took place at Alexandra Palace in London. It was warmly welcomed by the invited press with comments such as 'The new image builds on traditional values but adds a mood of optimism' and 'It isn't until you see the new livery and logo that it occurs to you that the old familiar one has become rather stale,' being offered in support of the changes.

A further identity change was revealed in spring 2008, when a new corporate scheme for all the UK coach, bus and rail operations was revealed. The Caetano Levanté, available exclusively to National Express and

The new National Express head office and coach station in Birmingham. *National Express*

its contracted operators, uses a new 'magic-floor' passenger lift to bring passengers in wheelchairs from the main door to a dedicated space at the front of the coach. This model now makes up the majority of the fleet although there are a few Plaxton Elites to the same specification. All new coaches now have leather seats and at-seat power sockets. Nearly 370 new coaches have entered service over the last three years and by early 2013 all coaches should be fully accessible.

Future developments

All coaches are now fitted with the Traffilog and UTrack systems. These enable the way in which the coaches are driven to be continuously monitored as well as tracking where they all are in real time and will shortly transfer this information to customer information screens on stations. Further developments will enable customers to track their journeys online and by using Smartphone apps.

Group Growth

National Express UK Ltd is the UK coach division of National Express Group which now has its global headquarters over the coach station in Birmingham. The business operates two train franchises, the UK's number one coach company, an airport and commuter shuttle bus services and bus companies in the West Midlands and Dundee. Full details of these bus fleet are included in the 'English Bus Handbook : Groups book'.

NATIONAL EXPRESS

National Express Ltd, National Express House, Birmingham Coach Station, Mill Lane, Birmingham, B5 6DD

A3	London - Gatwick Airport		CY
A6	London - Stansted Airport		SH
A9	London - Stansted Airport		SH
025	London - Gatwick Airport - Brighton		CY
026	London - Bognor Regis		CY
035	London - Bournemouth		SR
210	Heathrow Airport - Birmingham		SR
230	Heathrow Airport - Nottingham		SR
410	London - Wolverhampton		SR
420	London - Birmingham		SR
440	London - Leicester		SR
509	London - Cardiff		SR
701	Woking - Heathrow Airport		SR
727	Norwich - Gatwick Airport		CY
747	Brighton - Heathrow Airport		SR/C
787	Cambridge - Heathrow Airport		SR
797	Cambridge - Gatwick Airport		CY

1	SH	FJ56PFE	Scania K340 EB4			Caetano Levanté	C49FT	2006		
2	SH	FJ56PFF	Scania K340 EB4			Caetano Levanté	C49FT	2006		

3-21			Scania K340 EB6			Caetano Levanté	C61FT	2007		
3	SH	FJ07DVH	8	SH	FJ07DVO	13	SH	FN07BYX	18	SH	FJ57KHX
4	SH	FJ07DVK	9	SH	FJ07DVP	14	SH	FN07BYZ	19	SH	FJ57KHY
5	SH	FJ07DVL	10	SH	FJ07DVR	15	SH	FJ07BZA	20	SH	FJ57KHZ
6	SH	FJ07DVM	11	SH	FN07BYV	16	SH	FJ07BZB	21	SH	FJ57KJA
7	SH	FJ07DVN	12	SH	FN07BYW	17	SH	FJ07BZC			

22	SH	FJ57KHM	Scania K340 EB4			Caetano Levanté	C49FT	2007		
23	SH	FJ57KHK	Scania K340 EB4			Caetano Levanté	C49FT	2007		

24-39			Scania K340 EB6			Caetano Levanté	C61FT	2007		
24	SH	FJ57KHL	32	SH	FJ57KGZ	36	SH	FJ57KHD	38	SH	FJ57KHF
25	SH	FJ57KJE	33	SH	FJ57KHA	37	SH	FJ57KHE	39	SH	FJ57KHG
26	SH	FJ57KHT	35	SH	FJ57KHC						

41	SR	SP04HRM	Volvo B12M			TransBus Panther	C49FT	2004	Travel Tayside, 2006
45	SR	LK53KVW	Volvo B12B			Plaxton Panther	C49FT	2003	
46	SR	LK53KVT	Volvo B12B			Plaxton Panther	C49FT	2003	
48	SR	LK53KVP	Volvo B12B			Plaxton Panther	C49FT	2003	
49	SR	FH05URN	Volvo B12B			Caetano Enigma	C49FT	2005	
50	SR	FH05URR	Volvo B12B			Caetano Enigma	C49FT	2005	
51	SR	FJ57KJU	Scania K340 EB4			Caetano Levanté	C49FT	2007	
52	SR	FJ57KJO	Scania K340 EB4			Caetano Levanté	C49FT	2007	
53	SH	FJ57KHO	Scania K340 EB4			Caetano Levanté	C49FT	2007	
58	SH	YN55NDZ	Scania K114EB4			Irizar PB	C49FT	2006	

60-72			Volvo B12B			Caetano Levanté	C49FT	2006-07		
60	SR	FJ06URH	64	CY	FH06EAW	67	CY	FH06EAX	70	CY	FH06EBL
61	CY	FJ06GGK	65	SR	FN06FLH	68	CY	FH06EBM	71	CY	FJ06URO
62	CY	FJ06URG	66	CY	FH06FMA	69	CY	FH06EBN	72	CY	FJ07DWN
63	CY	FN06FMC									

74	SR	FH05URO	Volvo B12B			Caetano Enigma	C49FT	2005	
75	SR	FH05URP	Volvo B12B			Caetano Enigma	C49FT	2005	

16   The National Express Handbook

Depot code SH prefixes fleet number 142 on new arrival FJ12FYO, one of six Volvo B9Rs recently added to the fleet. It is seen in Stratford while heading towards London on service A9. *Dave Heath*

78-84			Volvo B12B		TransBus Panther			C49FT	2003		
78	CY	LK53KWE	80	CY	LK53KWB	83	CY	LK53KVY	84	CY	LK53KVX
79	CY	LK53KWD	82	CY	LK53KVZ						
95	SR	FJ57KHP	Scania K340 EB4		Caetano Levanté			C49FT	2008		
96	SR	FJ57KHR	Scania K340 EB4		Caetano Levanté			C49FT	2008		
102	SR	FJ10EZV	Scania K340 EB4		Caetano Levanté			C49FT	2010		
110-135			Volvo B9R		Caetano Levanté			C48FT	2010-11		
110	CY	FJ60HXS	117	SR	FJ11MKL	124	CY	FJ11MKF	130	CY	FJ11MJK
111	CY	FJ60HXT	118	SR	FJ11MKE	125	CY	FJ11MKG	131	CY	FJ11MJO
112	CY	FJ60HXU	119	SR	FJ11MKK	126	CY	FJ11MKM	132	CY	FJ11MJU
113	CY	FJ60HXV	120	SR	FJ11RDO	127	CY	FJ11MKN	133	CY	FJ11MJV
114	CY	FJ60HXX	121	SR	FJ11RDU	128	CY	FJ11MKU	134	CY	FJ11MJX
115	CY	FJ60HXY	122	SR	FJ11RDV	129	CY	FJ11MKV	135	CY	FJ11MJY
116	SR	FJ60KVS	123	CY	FJ11MKD						
136-143			Volvo B9R		Caetano Levanté			C48FT	2012		
136	SR	FJ12FYL	138	SR	FJ12FYN	140	SR	FJ12FYR	142	SR	FJ12FYO
137	SR	FJ12FYM	139	SR	FJ12FYP	141	SR	FJ12FYH	143	SR	FJ12FXD
A801	WD	X162ENJ	Mercedes-Benz Sprinter 311D		Frank Guy			M8	2001		
A802	WD	X164ENJ	Mercedes-Benz Sprinter 311D		Frank Guy			M8	2001		
A803	WD	Y228NLF	Mercedes-Benz Sprinter 311D		Frank Guy			M8	2002		
950	t	Y301HUA	DAF SB3000		Van Hool T9 Alizée			C49FT	2001		
951	t	Y302HUA	DAF SB3000		Van Hool T9 Alizée			C49FT	2001		
952	t	Y303HUA	DAF SB3000		Van Hool T9 Alizée			C49FT	2001		

The National Express Handbook

For a while coach 67, FH06EAX, also carried the St George flag. It was pictured in Norwich bus station while working the University of East Anglia to Gatwick service. *Mark Doggett*

8321-8352			ADL Dart 4 9.5m			ADL Enviro 200		N22F	2008		
8321	WD	SN08AAU	8329	WD	SN08ABU	8337	WD	SN08ACV	8345	WD	SN08ADZ
8322	WD	SN08AAV	8330	WD	SN08ABV	8338	WD	SN08ACX	8346	WD	SN08AEA
8323	WD	SN08AAX	8331	WD	SN08ABX	8339	WD	SN08ACY	8347	WD	SN08AEB
8324	WD	SN08AAY	8332	WD	SN08ABZ	8340	WD	SN08ACZ	8348	WD	SN08AEC
8325	WD	SN08AAZ	8333	WD	SN08ACF	8341	WD	SN08ADO	8349	WD	SN08AED
8326	WD	SN08ABF	8334	WD	SN08ACJ	8342	WD	SN08ADU	8350	CY	SN08AEE
8327	WD	SN08ABK	8335	WD	SN08ACO	8343	WD	SN08ADV	8351	CY	SN08AEF
8328	WD	SN08ABO	8336	WD	SN08ACU	8344	WD	SN08ADX	8352	CY	SN08AEG
8353	WD	T503TOL	Volvo B6BLE			Wright Crusader		N35F	1999		
8355-8358			ADL E20D			ADL Enviro 200		N29F	2011		
8355	WD	MX61BBF	8356	WD	MX61BBJ	8357	WD	MX61BBK	8358	WD	MX61BBN
8541-8547			Mercedes-Benz Citaro O530					AN30D	2008		
8541	WD	KX58GUA	8543	WD	KX58GUD	8545	WD	KX58GUF	8547	WD	KX58GUH
8542	WD	KX58GUC	8544	WD	KX58GUE	8546	WD	KX58GUG			
8548-8553			Mercedes-Benz Citaro O530					AN30D	2003	Quality Line, 2008	
8548	WD	BU53AXN	8550	WD	BU53AXP	8552	WD	BU53AXT	8553	WD	BU53AXV
8549	WD	BU53AXO	8551	WD	BU53AXR						
8554-8560			Mercedes-Benz Citaro O530 LE					N21	2008		
8554	WD	KX58GCJ	8556	WD	KX58GCL	8558	WD	KX58GTU	8560	WD	KX58GTZ
8555	WD	KX58GCK	8557	WD	KX58GCM	8559	WD	KX58GTY			
8561	WD	KX58GUJ	ADL Dart 4			ADL Enviro 200		N25F	2008		
8562	WD	KX58GUK	ADL Dart 4			ADL Enviro 200		N25F	2008		
8563	WD	KX58GUO	ADL Dart 4			ADL Enviro 200		N25F	2008		
8564-8573			Mercedes-Benz Sprinter 515cdi			KVC		M9	2009		
8564	WD	KX58BJK	8567	WD	KX58BJO	8570	WD	KX58BFA	8572	WD	KX09CJO
8565	WD	KX58BJV	8568	WD	KX58BKA	8571	WD	KX58BJY	8573	WD	KX09CJU
8566	WD	KX58BJU	8569	WD	KX58BJZ						
8574-8579			Mercedes-Benz Citaro O530					AN30D	2003-04	Arriva London, 2011	
8574	WD	BX04MXR	8576	WD	LX03HCG	8578	WD	LX03HCU	8579	WD	LX03HDE
8575	WD	LX03HCE	8577	WD	LX03HCL						

18   The National Express Handbook

National Express operates several services from West Drayton depot that support Heathrow airport, including the Hotel Hoppa network. These use Enviro 200 buses represented here by SN08ABO.
Mark Lyons

8600-8605   NAW Cobus 2700s   Cobus   N15T

8600	WD	8600	8602	WD	8602	8604	WD	8604	8605	WD	8605
8601	WD	8601	8603	WD	8603						

8606-8622   DAF SB220   East Lancs Myllennium   N29D   2000   Aviation Defence, 2009

8606	WD	X831NWX	8611	WD	X836NWX	8615	WD	X840NWX	8619	WD	X844NWX
8607	WD	X832NWX	8612	WD	X837NWX	8616	WD	X841NWX	8620	WD	X845NWX
8608	WD	X833NWX	8613	WD	X838NWX	8617	WD	X842NWX	8621	WD	X846NWX
8609	WD	X834NWX	8614	WD	X839NWX	8618	WD	X843NWX	8622	WD	X847NWX
8610	WD	X835NWX									

8623	WD	R425AOR	Denis Dart SLF	UVG UrbanStar	N25D	1998	Aviation Defence, 2009
8624	WD	R426AOR	Denis Dart SLF	UVG UrbanStar	N25D	1998	Aviation Defence, 2009

8625-8629   Volvo B6LE   Wrightbus Crusader   N21F   1996   Aviation Defence, 2009

8625	WD	N241WRW	8627	WD	N244WRW	8628	WD	N245WRW	8629	WD	N246WRW
8626	WD	N243WRW									

8637-8640   MAN 14.220   MCV   N28D   2006   Aviation Defence, 2009

8637	WD	AE55MVF	8638	WD	AE55MVG	8639	WD	AE55MVH	8640	WD	AE55MVJ

8641-8645   NAW Cobus 2700s   Cobus   N15T   2009

8641	WD	8641	8643	WD	8643	8644	WD	8645	8645	WD	8645
8642	WD	8642									

8646-8657   Mercedes-Benz Citaro O530   AN30D   2010

8646	WD	BK10EHT	8649	WD	BK10EHW	8652	WD	BK10EHZ	8655	WD	BK10EJX
8647	WD	BK10EHU	8650	WD	BK10EHX	8653	WD	BK10EJU	8656	WD	BK10EJY
8648	WD	BK10EHV	8651	WD	BK10EHY	8654	WD	BK10EJV	8657	WD	BK10EJZ

Depots and codes: Crawley (Whetstone Close, Tinsley Green) - CY; Bishops Stortford (Start Hill, Great Hollingbury) - SH and West Drayton (Sipson Road) - WD (buses) and SR (coaches). Details of the other National Express fleets may be found in the English Bus Handbook : Groups Handbook.

The National Express Handbook

ACE TRAVEL

A McLean, 3 Rockcliffe Park, Chapelhall, Airdrie, ML6 8SH

591	London - Edinburgh
596	London - Edinburgh

*	NX03MCL	Bova Futura FHD12.340	Bova	C53FT	2004	Bruce's, Salisburgh, '10
*	NX04MCL	Bova Futura FHD12.340	Bova	C53FT	2004	Bruce's, Salisburgh, '10
*	DX04MCL	Volvo B12B	Van Hool T9 Alizee	C49FT	2004	

Previous registrations:

NX03MCL	SF04WMY		DX04MCL	SP04AKG
NX04MCL	SF04WMX			

No vehicles are contracted in National Express colours. The vehicles used on the services are selected from the main fleet.

AMBASSADOR TRAVEL

Ambassador Travel (Anglia) Ltd, James Watt Close, Gapton Hall Industrial Estate, Great Yarmouth, NR31 0NX

308	Great Yarmouth - Birmingham
491	London - Lowestoft
497	London - Great Yarmouth

180	FD54DGU	Volvo B12B	VDL Jonckheere Mistral 50	C49FT	2005
181	FD54DGV	Volvo B12B	VDL Jonckheere Mistral 50	C49FT	2005
182	FD54DHX	Volvo B12B	VDL Jonckheere Mistral 50	C49FT	2005
183	FD54DHY	Volvo B12B	VDL Jonckheere Mistral 50	C49FT	2005
206	FJ09DXA	Scania K340 EB4	Caetano Levanté	C49FT	2009
207	FJ09DXB	Scania K340 EB4	Caetano Levanté	C49FT	2009
208	FJ09DXC	Scania K340 EB4	Caetano Levanté	C49FT	2009
209	FJ09DXE	Scania K340 EB4	Caetano Levanté	C49FT	2009

Details of other vehicles in this fleet may be found in the *English Bus Handbook: Coaches* book.

BAYLISS EXECUTIVE TRAVEL

Unit 1, Minters Industrial Estate, Southwall Road, Deal, Kent CT14 9PZ

650	London - Dover (Cruise Terminal)

No vehicles are contracted in National Express colours. The vehicles used on the services are selected from the main fleet.

Turning out of Elizabeth Street en route for Cambridge is FJ09DXE, one of four Scania K340s operated by Ambassador Travel. Arriva subsidiary TGM currently provide the vehicle for route 010. *Mark Bailey*

BENNETT'S

Bennetts Coaches, Eastern Avenue, Gloucester, GL4 4LP

222   Gatwick Airport - Heathrow Airport - Gloucester

BE1	FJ60EFP	Volvo B9R	Caetano Levanté	C48FT	2010
BE2	FJ60EFR	Volvo B9R	Caetano Levanté	C48FT	2010
BE3	FJ60EFS	Volvo B9R	Caetano Levanté	C48FT	2010
BE4	FJ60EFT	Volvo B9R	Caetano Levanté	C48FT	2010
*	FJ60KUH	Mercedes-Benz OC500RF	Caetano Levanté	C48FT	2010
*	FJ60KUK	Mercedes-Benz OC500RF	Caetano Levanté	C48FT	2010

BL TRAVEL

B Lockwood, The Garage, Hoyle Mill Road, Kinsley, Wakefield WF9 5JB

561   London - Bradford

No vehicles are contracted in National Express colours. The vehicles used on the services are selected from the main fleet.

The National Express Handbook

BRUCE'S

J Bruce, 40 Main Street, Salsburgh, ML7 4LA

336	Edinburgh - Plymouth	
532	Edinburgh - Plymouth	
539	Edinburgh - Bournemouth	
588	Inverness - London	
590	Aberdeen - Glasgow - London	

FJ57KGE	Scania K340 EB6	Caetano Levanté	C61FT	2007	National Express, 2010
FJ57KGY	Scania K340 EB6	Caetano Levanté	C61FT	2007	National Express, 2009
FJ57KHB	Scania K340 EB6	Caetano Levanté	C61FT	2007	National Express, 2009
FJ57KHU	Scania K340 EB6	Caetano Levanté	C61FT	2007	National Express, 2009
FJ57KHV	Scania K340 EB6	Caetano Levanté	C61FT	2007	National Express, 2009
FJ57KHW	Scania K340 EB6	Caetano Levanté	C61FT	2007	National Express, 2009
FJ57KJF	Scania K340 EB6	Caetano Levanté	C61FT	2007	National Express, 2009
FJ58AKK	Scania K340 EB6	Caetano Levanté	C61FT	2008	
FJ58AKY	Scania K340 EB6	Caetano Levanté	C61FT	2008	
YV12CZJ	Scania K340 EB6	Caetano Levanté	C61FT	2012	

Details of the other vehicles in this fleet may be found in the *Scottish Bus Handbook*.

BURGIN'S EUROPEAN

Unit 7, 35 Catley Road, Sheffield, S9 5JF

320	Bradford - Birmingham
561	Bradford - London

No vehicles are contracted in National Express colours. The vehicles used on the service are selected from the main fleet.

BUZZLINES TRAVEL

Busslines Travel, Lympne Distribution Park, Lympne, Hythe, CT21 4LR

650	London - Dover (Cruise Terminal)

No vehicles are contracted in National Express colours. The vehicles used on the service are selected from the main fleet.

CHALFONT

Chalfont Coaches of Harrow Ltd, 200 Featherstone Road, Southall, UB2 5AQ

024	London - Eastbourne
035	London - Bournemouth
040	London - Bristol
450	London - Nottingham
460	London - Stratford-upon-Avon
502	London - Ilfracombe
509	London - Cardiff
560	London - Sheffield

CD1	WA59EBC	Volvo B12B	Van Hool T9 Acron	C53FT	2009
CD2	WA10ENL	Volvo B12B	Van Hool T9 Acron	C53FT	2010
CD3	WA61AKP	Volvo B12B	Van Hool T9 Acron	C53FT	2011
CD4	WA61AKU	Volvo B12B	Van Hool T9 Acron	C53FT	2011
CD5	FJ61EWN	Volvo B9R	Caetano Levanté	C48FT	2011
CD6	FJ61EWO	Volvo B9R	Caetano Levanté	C48FT	2011

Details of the vehicles in this fleet may be found in the *English Bus Handbook : Coaches* book

Just eleven Van Hool T9 coaches are on the Approved Vehicle list of National Express coaches. The four with Chalfont feature the tri-axle B12B chassis. WA59EBC was working the Cardiff service when pictured.
Mark Bailey

R W CHENERY

PG Garnham, The Garage, Dickleburgh, Diss, Norfolk, IP21 4NJ

490	London - Norwich				
	FJ05AOV	Volvo B12B	VDL Jonckheere Mistral	C49FT	2005
	N999RWC	Volvo B12B	VDL Jonckheere Mistral	C49FT	2005
	FJ05AOX	Volvo B12B	VDL Jonckheere Mistral	C49FT	2005
	FJ11MKA	Volvo B9R	Caetano Levanté	C48FT	2011
	FJ11MKC	Volvo B9R	Caetano Levanté	C48FT	2011

Previous registration:
N999RWC   FJ05AOW

Details of the vehicles in this fleet may be found in our *English Bus Handbook : Coaches* book

The Chenery contribution to National Express work includes three VDL Jonckheere Mistral coaches represented here by FJ05AOX. It is seen at the northern end of the service from Norwich to London.
Mark Doggett

EAST YORKSHIRE

East Yorkshire Motor Services Ltd, 252 Anlaby Road, Hull, HU3 2RS

031	London - Portsmouth
322	Hull - Swansea
327	Scarborough - Bath
562	London - Hull

62	YX07HJD	Volvo B12B	Caetano Levanté	C49FT	2007
63	YX07HJE	Volvo B12B	Caetano Levanté	C49FT	2007
64	YX07HJF	Volvo B12B	Caetano Levanté	C49FT	2007
65	YX07HJG	Volvo B12B	Caetano Levanté	C49FT	2007
66	YX07HJJ	Volvo B12B	Caetano Levanté	C49FT	2007
67	YX07HJK	Volvo B12B	Caetano Levanté	C49FT	2007
68	YX08FYP	Volvo B12B	Caetano Levanté	C49FT	2008

Details of the other vehicles in this fleet may be found in the *English Bus Handbook : Smaller Groups* book.

A long time contributor to National Express work is East Yorkshire whose contribution comprises entirly Caetano Levanté-bodied Volvo B12Bs. Illustrating the type is the latest example, 68, YX08FYP.
Mark Bailey

The National Express Handbook 25

EDWARDS COACHES

Edwards Coaches Ltd, Newtown Ind Est, Llantwit Fardre, Pontypridd CF38 2EE

201	Swansea - Gatwick Airport	
202	Carfidd - Heathrow Airport	
320	Cardiff - Bradford	
321	Aberdare - Bradford	
322	Swansea - Hull	
507	Swansea - London	
508	Haverfordwest - London	
509	London - Swansea - Cardiff	
528	Haverfordwest - Rochdale	

ED1	YN08DNV	Volvo B12B		Plaxton Panther		C49Ft	2008	
ED2-12		Volvo B9R		Caetano Levanté		C48FT	2011	
ED2	FJ11GNK	ED5	FJ11GKY	ED8	FJ11GNN		ED11	FJ11GMX
ED3	FJ11GNO	ED6	FJ11GMZ	ED9	FJ11GMY		ED12	FJ11GMO
ED4	FJ11GNF	ED7	FJ11GOH	ED10	FJ11GMU			
ED13-33		Volvo B9R		Caetano Levanté		C48FT	2012	
ED13	FJ12FXC	ED19	FJ12FXU	ED24	FJ12FXZ		ED29	FJ12FYE
ED14	FJ12FXM	ED20	FJ12FXV	ED25	FJ12FYA		ED30	FJ12FYF
ED15	FJ12FXO	ED21	FJ12FXW	ED26	FJ12FYB		ED31	FJ12FYG
ED16	FJ12FXP	ED22	FJ12FXX	ED27	FJ12FYC		ED32	FJ12FYH
ED17	FJ12FXR	ED23	FJ12FXY	ED28	FJ12FYD		ED33	FJ12FYK
ED18	FJ12FXS							

Details of the other vehicles in this fleet may be found in the *Welsh Bus Handbook*.

Edwards coaches is now responsible for many of the services originating in south Wales that latterly First undertook. From the fleet of thirty-three dedicated vehicles, ED12, FJ11GMO is shown. *Mark Bailey*

ELCOCK REISEN

M H Elcock & Son Limited, 66 High Street, Dawley, Telford, TF4 2HD

675		Wolverhampton - Minehead (Butlins)				
*	3419NT	Volvo B12B	Van Hool T9 Alizee	C49FT	2005	Nyfen Coaches, 2006

Previous registration:
3419NT   CX05AHU

No vehicles are contracted in National Express. Details of the other vehicles in this fleet may be found in the *English Bus Handbook: Coaches* book.

EPSOM COACHES - RAPT

Epsom Coaches, Roy Richmond Way, Epsom, KT19 9AF

410		London - Birmingham			
550		London - Liverpool			
EP1	FJ11GLF	Volvo B9R	Caetano Levanté	C48FT	2011
EP2	FJ11GMV	Volvo B9R	Caetano Levanté	C48FT	2011
EP3	FJ61EYK	Volvo B9R	Caetano Levanté	C48FT	2012
EP4	FJ61EYL	Volvo B9R	Caetano Levanté	C48FT	2012

Details of the other vehicles in this fleet may be found in the *English Bus Handbook: Coaches* book

Epsom Coaches, now part of the French RAPT Group, operates duties on routes 410 and 550. Seen passing along Park Lane in London is EP2, FJ11GMV. *Dave Heath*

EXCELSIOR

Excelsior Coaches Ltd, Central Business Park, Southcote Road, Bournemouth, BH1 3SJ

035	Bournemouth University - London	
205	Poole - Gatwick Airport	
206	Poole - Gatwick Airport	
310	Poole - Bradford	

909	A17XEL	Volvo B12B	Caetano Enigma	C49FT	2006
910	A18XEL	Volvo B12B	Caetano Enigma	C49FT	2006
911	A19XEL	Volvo B12B	Caetano Enigma	C49FT	2006
912	A20XEL	Volvo B12B	Caetano Enigma	C49FT	2006
913	BU08CGG	Volvo B12B	Caetano Enigma	C49FT	2008
914	FJ60EHB	Volvo B9R	Caetano Levante	C48FT	2010
915	FJ60EHC	Volvo B9R	Caetano Levante	C48FT	2010
916	FJ60EHD	Volvo B9R	Caetano Levante	C48FT	2010
917	FJ60EHE	Volvo B9R	Caetano Levante	C48FT	2010
918	FJ60EHF	Volvo B9R	Caetano Levante	C48FT	2010
919	FJ11GNP	Volvo B9R	Caetano Levante	C48FT	2011
920	FJ11GNU	Volvo B9R	Caetano Levante	C48FT	2011
921	FJ11GNV	Volvo B9R	Caetano Levante	C48FT	2011
922	FJ11GNX	Volvo B9R	Caetano Levante	C48FT	2011
923	FJ11GNY	Volvo B9R	Caetano Levante	C48FT	2011
924	FJ11GNZ	Volvo B9R	Caetano Levante	C48FT	2011

Previous registrations:

A17XEL	FJ55BXZ		A19XEL	FJ55BXW
A18XEL	FJ55BXV		A20XEL	FJ55BXY

Details of the other vehicles in this fleet may be found in the English Bus Handbook: Coaches book

GALLOWAY

Galloway European Coachlines Ltd, Denter's Hill, Mendlesham, Stowmarket, IP14 5RR

250	Ipswich - Heathrow Airport	
481	London - Felixstowe	

220	2513PP	VDL Bus SB4000	Van Hool T9 Alizée	C49FT	2004
221	2086PP	VDL Bus SB4000	Van Hool T9 Alizée	C49FT	2004

300-305   Volvo B9R   Caetano Levanté   C48FT   2011

300	FJ61EVN	302	FJ61EVR	304	FJ61EVU	305	FJ61EVV
301	FJ61EVP	303	FJ61EVT				

Previous registrations:

2086PP   YJ54CJE   2513PP   YJ54CKN

Details of the vehicles in this fleet may be found in the *English Bus Handbook : Coaches* book

Two Volvo B9Rs with Caetano Levanté bodywork are shown. Above is 919, FJ11GNP operated by Excelsior of Bournemouth while below 301, FJ61EVP, from the Galloway fleet is seen operating route 250. As we go to press the number of Volvo B9Rs on the network is two hundred and fifty with a further forty-eight with Plaxton Elite bodywork. *Mark Doggett*

GO NORTH EAST

Go North East Ltd, 117 Queen Street, Bensham, Gateshead, NE8 2UA

326	Newcastle-upon-Tyne - Cambridge				
332	Newcastle-upon-Tyne - Birmingham - Gloucester				
380	Newcastle-upon-Tyne - Liverpool				
530	Paignton - Newcastle-upon-Tyne				
531	Plymouth - Newcastle-upon-Tyne				
580	Newcastle-upon-Tyne - Liverpool				
663	Newcaste-upon-Tyne - Skegness				

7090	CU6860	Volvo B12B	TransBus Panther	C49FT	2004
7092	K2VOY	Volvo B12B	Plaxton Panther	C49FT	2006
7093	K3VOY	Volvo B12B	Plaxton Panther	C49FT	2006
7094	FJ08KLF	Scania K340 EB6	Caetano Levanté	C53FT	2008
7095	FJ08KLS	Scania K340 EB6	Caetano Levanté	C53FT	2008
7096	FJ08KLU	Scania K340 EB6	Caetano Levanté	C53FT	2008
7097	FJ08KLX	Scania K340 EB6	Caetano Levanté	C53FT	2008
7098	FJ08KLZ	Scania K340 EB6	Caetano Levanté	C53FT	2008
7099	FJ08KMU	Scania K340 EB6	Caetano Levanté	C53FT	2008
7100	FJ08KMV	Scania K340 EB6	Caetano Levanté	C53FT	2008
7101	FJ08KNV	Scania K340 EB6	Caetano Levanté	C53FT	2008
7102	FJ08KNW	Scania K340 EB6	Caetano Levanté	C53FT	2008
7103	JCN822	Volvo B9R	Caetano Levanté	C48FT	2012
7104	574CPT	Volvo B9R	Caetano Levanté	C48FT	2012

Previous registrations:

574CPT	FJ61GZD	JCN822	FJ61GZC

Details of the vehicles in this fleet may be found in the annual *Go-Ahead Bus Handbook*.

Seen near its home base in Gateshead is Go-Ahead Northern's 7097, FJ08KLX, a Scania K340 with Caetano Levanté bodywork. It was operating route 380 from Liverpool when pictured. Go North East provides two journeys each way on this service. *Mark Bailey*

GO SOUTH COAST

Wilts & Dorset Bus Co. Ltd, Towngate House, 2-8 Parkstone Road, Poole, BH15 2PR

033   London - Salisbury - Yeovil

7051	FX61EVX	Volvo B9R	Caetano Levanté	C48FT	2011
7131	YN07EWS	Scania K340 EB4	Caetano Levanté	C49FT	2007
7132	YN07EWT	Scania K340 EB4	Caetano Levanté	C49FT	2007

Details of the other vehicles in this fleet may be found in the annual *Go-Ahead Bus Handbook*.

Part of Go-Ahead's South Coast operation, Wilts & Dorset, provide three journeys each way between London and south west England, two of which work through to Yeovil. Seen leaving London for Salisbury on the third journey of the day is 7131, YN07EWS. *Mark Bailey*

The National Express Handbook

JOHNSON BROS

Johnsons Bros, Green Acres, Green Lane, Hodthorpe, Worksop, S80 4XR

350		Liverpool - Mansfield - Cambridge			
JOH1	FJ60EGE	Volvo B9R	Caetano Levanté	C48FT	2010
JOH2	FJ60EGF	Volvo B9R	Caetano Levanté	C48FT	2010

Details of the vehicles in this fleet may be found in the *English Bus Handbook : Coaches* book

THE KINGS FERRY

The Kings Ferry Coach Company, Eastcourt Lane, Gillingham, ME8 6HW

007		London - Dover			
6.17	FJ10EZP	Mercedes-Benz OC500	Caetano Levanté	C53FT	2010
6.18	FJ10EZR	Mercedes-Benz OC500	Caetano Levanté	C53FT	2010

Details of the vehicles in this fleet may be found in the *English Bus Handbook : Smaller Groups* book

Kings Ferry is part of the National Express group and its full fleet can be found in the National Express section of our *English Bus Handbook : Smaller Groups* book. Two coaches are currently dedicated to National Express sevice, both Mercedes-Benz OC500s with Caetano Levanté bodywork, the only examples of this chassi used on the network. A similar vehicle, FJ10EZM, carries fleet colours. *Colin Lloyd*

LLEW JONES

Llew Jones International, Station Yard, Station Rd, Llanrwst, LL26 0EH

| 385 | Bangor - Manchester |
| 396 | Pwllheli - Manchester |

LJ12LLJ	Volvo B9R	Caetano Levanté	C48FT	2012

Other vehicles used on the service are selected from the main fleet.

Displaying rear promotion for the Portsmouth service is FJ58AHL from Luckett's fleet.
Dave Heath

A comparison may be made between the grill on the Scania above and the Volvo B9R here. FJ61EVW is seen working route 203.
Mark Bailey

LUCKETTS

H Luckett & Co Ltd, Broadcut, Wallington, Fareham, PO16 8TB

030	London - Fareham
203	Heathrow Airport - Portsmouth - Southsea
300	Bristol - Southsea
654	London - Portsmouth Port
668	London - Bodnor Regis (Butlins)

X4804	FJ11RDX	Volvo B9R	Caetano Levanté	C48FT	2011
X4805	FJ11RDY	Volvo B9R	Caetano Levanté	C48FT	2011
X4806	FJ61EWA	Volvo B9R	Caetano Levanté	C48FT	2011
X4807	FJ61EVW	Volvo B9R	Caetano Levanté	C48FT	2011
X4951	FJ58AHE	Scania K340 EB4	Caetano Levanté	C49FT	2009
X4952	FJ58AHF	Scania K340 EB4	Caetano Levanté	C49FT	2009
X4953	FJ58AHG	Scania K340 EB4	Caetano Levanté	C49FT	2009
X4954	FJ58AHK	Scania K340 EB4	Caetano Levanté	C49FT	2009
X4955	FJ58AHL	Scania K340 EB4	Caetano Levanté	C49FT	2009
X4956	FJ58AJY	Scania K340 EB4	Caetano Levanté	C49FT	2009
X4957	FJ58AKF	Scania K340 EB4	Caetano Levanté	C49FT	2009
X4958	FJ58AKG	Scania K340 EB4	Caetano Levanté	C49FT	2009
X4959	FJ58AHN	Scania K340 EB4	Caetano Levanté	C49FT	2009
X4960	FJ58AJX	Scania K340 EB4	Caetano Levanté	C49FT	2009
X4961	FJ58AHO	Scania K340 EB4	Caetano Levanté	C49FT	2009
X4962	FJ58AHP	Scania K340 EB4	Caetano Levanté	C49FT	2009
X4965	FJ59AOZ	Scania K340 EB4	Caetano Levanté	C49FT	2010
X4966	FJ59APF	Scania K340 EB4	Caetano Levanté	C49FT	2010

Details of the vehicles in this fleet may be found in the *English Bus Handbook : Coaches* book

MAINLINE

Mainline Coaches Ltd, Kings Head Garage, Glannant Road, Evanstown, CF39 8RL

| 509 | London - Cardiff |
| 672 | Swansea - Minehead |

No vehicles are contracted in National Express colours. The vehicles used on the services are selected from the main fleet.

MIKE De COURCEY

Mike de Courcey Travel Ltd, Rowley Drive, Coventry CV3 4FG

210	Birmingham - Gatwick Airport
310	Coventry - Leeds
325	Birmingham - Manchester
384	Birmingham - Llandudno
387	Coventry - Blackpool
410	London - Birmingham - Wolverhampton
420	London - Birmingham - Wolverhampton
460	London - Stratford-upon-Avon - Coventry
540	London - Manchester
661	Coventry - Skegness

MD1-14   Volvo B9R   Caetano Levanté   C48FT   2011

MD1	FJ11MKZ	MD5	FJ11GME	MD9	FJ11GKZ	MD12	FJ11GKG
MD2	FJ11GKO	MD6	FJ11GKD	MD10	FJ11MLU	MD13	FJ11GKF
MD3	FJ11GMF	MD7	FJ11GKN	MD11	FJ11GMG	MD14	FJ11GKK
MD4	FJ11GKL	MD8	FJ11GJV				

Details of the other vehicles in this fleet may be found in the *English Bus Handbook : Notable Independents* book.

In 2011 Mike de Courcey Travel commenced National Express work from its base in Coventry, with fourteen Volvo B9R coaches. Seen at Hyde Park in London is MD5, FJ11GME. *Dave Heath*

The National Express Handbook

PARK'S OF HAMILTON

Parks of Hamilton (Coach Hirers) Ltd, 20 Bothwell Road, Hamilton, ML3 0AY

Walkham Park, Burrington Way, Plymouth, PL5 3LS

315	Helston - Eastbourne
328	Plymouth - Blackpool
404	London - Penzance
406	London - Newquay
421	London - Blackpool
500	London - Penzance
501	London - Plymouth - Totnes
504	London - Penzance
534	Glasgow - Hull
537	Glasgow - Corby
538	Inverness - Glasgow - Manchester Airport - Coventry
540	London - Bury - Rochdale
570	London - Blackpool
592	London - Glasgow - Aberdeen

KSK948	Volvo B12B 15m	Plaxton Panther	C65FT	2009
KSK949	Volvo B12B 15m	Plaxton Panther	C65FT	2009
LSK506	Volvo B12B 15m	Plaxton Panther	C65FT	2009
LSK507	Volvo B12B 15m	Plaxton Panther	C65FT	2008
LSK510	Volvo B12B 15m	Plaxton Panther	C65FT	2009
LSK511	Volvo B12B 15m	Plaxton Panther	C65FT	2009
LSK512	Volvo B12B 15m	Plaxton Panther	C65FT	2009
LSK513	Volvo B12B 15m	Plaxton Panther	C65FT	2009
LSK611	Volvo B12B 15m	Plaxton Panther	C65FT	2009
LSK613	Volvo B12B 15m	Plaxton Panther	C65FT	2009

Park's fleet all carry index numbers from a series supplied by the Scottish office and which are transferred to new vehicles as they arrive. The fleet used on National Express work is comprised entirely of Plaxton-bodied Volvo coaches with Panther bodywork shown on **LSK835**. *Dave Heath*

LSK830	Volvo B12B 15m	Plaxton Panther	C65FT	2009
LSK831	Volvo B12B 15m	Plaxton Panther	C65FT	2009
LSK832	Volvo B12B 15m	Plaxton Panther	C65FT	2009
LSK835	Volvo B12B 15m	Plaxton Panther	C65FT	2009
LSK839	Volvo B12B 15m	Plaxton Panther	C65FT	2009
LSK845	Volvo B12B 15m	Plaxton Panther	C65FT	2009
LSK870	Volvo B12B 15m	Plaxton Panther	C65FT	2009
HSK642	Volvo B9R 12.6m	Plaxton Elite	C48FT	2011
HSK643	Volvo B9R 12.6m	Plaxton Elite	C48FT	2011
HSK644	Volvo B9R 12.6m	Plaxton Elite	C48FT	2011
HSK645	Volvo B9R 12.6m	Plaxton Elite	C48FT	2011
HSK646	Volvo B9R 12.6m	Plaxton Elite	C48FT	2011
KSK950	Volvo B9R 12.6m	Plaxton Elite	C48FT	2011
KSK951	Volvo B9R 12.6m	Plaxton Elite	C48FT	2011
KSK952	Volvo B9R 12.6m	Plaxton Elite	C48FT	2011
KSK953	Volvo B9R 12.6m	Plaxton Elite	C48FT	2011
KSK986	Volvo B9R 12.6m	Plaxton Elite	C48FT	2011
HSK651	Volvo B9R	Plaxton Elite	C48FT	2012
HSK652	Volvo B9R	Plaxton Elite	C48FT	2012
HSK653	Volvo B9R	Plaxton Elite	C48FT	2012
HSK654	Volvo B9R	Plaxton Elite	C48FT	2012
HSK655	Volvo B9R	Plaxton Elite	C48FT	2012
HSK656	Volvo B9R	Plaxton Elite	C48FT	2012
HSK657	Volvo B9R	Plaxton Elite	C48FT	2012
HSK658	Volvo B9R	Plaxton Elite	C48FT	2012
HSK659	Volvo B9R	Plaxton Elite	C48FT	2012
HSK660	Volvo B9R	Plaxton Elite	C48FT	2012
KSK984	Volvo B9R	Plaxton Elite	C48FT	2012
KSK985	Volvo B9R	Plaxton Elite	C48FT	2012

Previous registrations:
LSK507   LSK511

Depots: Plymouth and Hamilton. Details of the other vehicles in this fleet may be found in *The Scottish Bus Handbook*.

The latest arrivals are further Plaxton Elite-bodied Volvo B9Rs. Pictured in Glasgow, KSK986 from the 2011 delivery is seen arriving on the service from Hull. *Mark Doggett*

PETER GODWARD COACHES

P Godward, 4 Edwin Hall View, South Woodham Ferrers, CM3 5QL

090	London - Southend-on-Sea					
540	London - Manchester					
550	London - Liverpool					
561	London - Leeds					
570	London - Whitehaven					

PG1	YN08MOA	Scania K340 EB4	Irizar Century	C49FT	2008	
PG2	FJ08KMX	Scania K340 EB4	Caetano Levanté	C49FT	2008	
PG3	MX56HZA	Scania K340 EB4	Caetano Levanté	C49FT	2006	Haytons, Manchester, 2012

PREMIER

Premier Travel Ltd, Trent Wharf, Meadow Lane, Nottingham, NG2 3HR

310	Leicester - Bradford
330	Nottingham - Penzance
397	Leicester - Blackpool
440	Leicester - London
561	London - Bradford

974	FJ61EWX	Volvo B9R	Caetano Levanté	C48FT	2011	
975	FJ61EWY	Volvo B9R	Caetano Levanté	C48FT	2011	
976	FJ59ARF	Scania K340 EB4	Caetano Levanté	C49FT	2009	Coachworks, Nottingham
977	FJ59APZ	Scania K340 EB4	Caetano Levanté	C49FT	2009	Coachworks, Nottingham
978	FJ61EYF	Volvo B9R	Caetano Levanté	C48FT	2012	
979	FJ61EYG	Volvo B9R	Caetano Levanté	C48FT	2012	
980	FJ61EYH	Volvo B9R	Caetano Levanté	C48FT	2012	

No vehicles are contracted in National Express colours. The vehicles used on the services are selected from the main fleet. Details of the vehicles in this fleet may be found in the *English Bus Handbook : Coaches* book

Peter Godward provides three vehicles on National Express work from its Essex base. Seen at the end of its journey from Cumbria is Scania K340 FJ08KMX. *Dave Heath*

Two coaches are required for the daily twelve-hour journey from Nottingham to Penzance which Premier commenced operating in 2011. Seen on the route is FJ59APZ, one of a pair of Scania K340s operated.
Mark Doggett

ROTALA

Flight Hallmark, Beacon House, Long Acre, Aston, Birmingham, B7 5JJ

| 210 | Gatwick Airport - Birmingham |
| 777 | Stansted Airport - Birmingham |

13101	FJ11GLK	Volvo B9R	Caetano Levanté	C48FT	2011
13102	FJ11GLZ	Volvo B9R	Caetano Levanté	C48FT	2011
13103	FJ11GKP	Volvo B9R	Caetano Levanté	C48FT	2011
13104	FJ11GKU	Volvo B9R	Caetano Levanté	C48FT	2011
13105	FJ11GKV	Volvo B9R	Caetano Levanté	C48FT	2011
13106	FJ11GKX	Volvo B9R	Caetano Levanté	C48FT	2011
13107	FJ11GJO	Volvo B9R	Caetano Levanté	C48FT	2011
13108	FJ11GJU	Volvo B9R	Caetano Levanté	C48FT	2011
13109	FJ11MLE	Volvo B9R	Caetano Levanté	C48FT	2011
13110	FJ11MLF	Volvo B9R	Caetano Levanté	C48FT	2011
13111	FJ11MLK	Volvo B9R	Caetano Levanté	C48FT	2011
13112	FJ11MLL	Volvo B9R	Caetano Levanté	C48FT	2011
13113	FJ11MLN	Volvo B9R	Caetano Levanté	C48FT	2011
13114	FJ11MLO	Volvo B9R	Caetano Levanté	C48FT	2011

Details of the other vehicles in this fleet may be found in our *English Bus Handbook: Smaller Groups* book

Rotala is one of the expanding smaller operators whose networks include Birmingham-based Flight Hallmark as well as Central Connect. In 2011 it started National Express operations from Birmingham to Gatwick and Stansted airports using fourteen coaches. FJ11MLE, now with fleet number 13109, is seen in Stevenage. *Dave Heath*

SELWYNS

Haytons - Selwyns

Selwyns Travel Ltd, Cavendish Farm Road, Weston, Runcorn, WA7 4LU

060		Liverpool - Manchester - Leeds		H/S		
061		Leeds - Manchester Airport		H/S		
304		Liverpool - Weymouth		S		
305		Liverpool - Southend		S		
314		Southport - Cambridge		S		
323		Liverpool - Cardiff		S		
325		Manchester - Birmingham		H		
350		Liverpool - Clacton		S		
381		Wrexham - Leeds		S		
410		London - Birminham		H		
417		London - Stafford		S		
418		London - Wrexham		S		
422		London - Burnley		S		
440		London - Manchester		H		
540		London - Bury - Burnley		H/S		
550		London - Birkenhead - Southport		S		
580		Liverpool - Newcastle-upon-Tyne		S		
109	YJ54CFD	VDL Bus SB4000	Van Hool T9 Alizée	C49FT	2004	
111	YJ05PWE	VDL Bus SB4000	Van Hool T9 Alizée	C49FT	2005	
112	YJ05PWF	VDL Bus SB4000	Van Hool T9 Alizée	C49FT	2005	
113	FH06EBO	Volvo B12B	Caetano Levanté	C49FT	2006	
114	FJ56PAO	Volvo B12B	Caetano Levanté	C49FT	2006	
115	FJ56PBF	Volvo B12B	Caetano Levanté	C49FT	2006	
116	FJ56PBO	Volvo B12B	Caetano Levanté	C49FT	2006	
127	YJ05PXF	VDL Bus SB4000	Van Hool T9 Alizée	C49FT	2005	City Circle, Kensington '08
128	YJ06LFV	VDL Bus SB4000	Van Hool T9 Alizée	C49FT	2006	City Circle, Kensington '08
131	FJ58AJU	Scania K340 EB4	Caetano Levanté	C49FT	2009	
132	FJ58AJV	Scania K340 EB4	Caetano Levanté	C49FT	2009	
139	FJ59APX	Scania K340 EB4	Caetano Levanté	C49FT	2009	
140	FJ59APY	Scania K340 EB4	Caetano Levanté	C49FT	2009	

Selwyn's recently acquired the Manchester business of Haytons which increased its activity on national Express services. Southampton is the location for this view of YN11AYC, a Plaxton-bodied Volvo B9R.
Dave Heath

Hanley bus station in Staffordshire provides the location for this view of Selwyns YN10FKP, a Plaxton Elite-bodied Volvo B9R. Route 550 links Southport with London taking in Liverpool and Birkenhead, both terminal points for some journeys. *Cliff Beeton*

141	YN10FKM	Volvo B9R	Plaxton Elite	C48FT	2010	
142	YN10FKO	Volvo B9R	Plaxton Elite	C48FT	2010	
143	YN10FKP	Volvo B9R	Plaxton Elite	C48FT	2010	
144	YN10FKR	Volvo B9R	Plaxton Elite	C48FT	2010	
145	YN10FKS	Volvo B9R	Plaxton Elite	C48FT	2010	
146	YN10FKT	Volvo B9R	Plaxton Elite	C48FT	2010	
147	YN10FKV	Volvo B9R	Plaxton Elite	C48FT	2010	
153	YN11AYA	Volvo B9R	Plaxton Elite	C48FT	2011	
154	YN11AYB	Volvo B9R	Plaxton Elite	C48FT	2011	
155	YN11AYC	Volvo B9R	Plaxton Elite	C48FT	2011	
156	YN11AYD	Volvo B9R	Plaxton Elite	C48FT	2011	
161	FJ11MLV	Volvo B9R	Caetano Levanté	C48FT	2011	
165	FJ61EWF	Volvo B9R	Caetano Levanté	C48FT	2011	
166	FJ61EWG	Volvo B9R	Caetano Levanté	C48FT	2011	
167	FJ61EWH	Volvo B9R	Caetano Levanté	C48FT	2011	
168	FJ61EWK	Volvo B9R	Caetano Levanté	C48FT	2011	
169	FJ61EWL	Volvo B9R	Caetano Levanté	C48FT	2011	
170	FJ61EWM	Volvo B9R	Caetano Levanté	C48FT	2011	
171	FJ61EWB	Volvo B9R	Caetano Levanté	C48FT	2011	
172	FJ61EWZ	Volvo B9R	Caetano Levanté	C48FT	2011	
173	FJ61EXK	Volvo B9R	Caetano Levanté	C48FT	2011	
174	FJ61EXL	Volvo B9R	Caetano Levanté	C48FT	2011	
182	YN55WSO	Volvo B12B	Plaxton Panther	C49FT	2005	Haytons, Manchester, 2011
186	FJ60EFU	Volvo B9R	Caetano Levanté	C48FT	2010	Haytons, Manchester, 2011
187	FJ60EFV	Volvo B9R	Caetano Levanté	C48FT	2010	Haytons, Manchester, 2011
188	FJ60EFW	Volvo B9R	Caetano Levanté	C48FT	2010	Haytons, Manchester, 2011

Details of the vehicles in this fleet may be found in our *English Bus Handbook : Coaches* book.

SILVERDALE

Silverdale Tours (Nottingham) Ltd, Little Tennis Street South, Nottingham, NG2 4EU

329	Nottingham - Newcastle-upon-Tyne				
440	London - Derby				
450	London - Mansfield - Retford				
767	Stansted Airport - Nottingham				
	FJ55DYW	Volvo B12B	Plaxton Panther	C49FT	2006
	FJ55DZK	Volvo B12B	Plaxton Panther	C49FT	2006
	FJ58AJO	Scania K340 EB4	Caetano Levanté	C49FT	2008
	FJ09DWY	Scania K340 EB4	Caetano Levanté	C49FT	2009
	FJ09DXT	Scania K340 EB4	Caetano Levanté	C49FT	2009
	FJ09DXU	Scania K340 EB4	Caetano Levanté	C49FT	2009
	FJ10EZT	Scania K340 EB4	Caetano Levanté	C49FT	2010
	FJ10EZU	Scania K340 EB4	Caetano Levanté	C49FT	2010
	FJ61EWP	Volvo B9R	Caetano Levanté	C48FT	2012
	FJ61EWR	Volvo B9R	Caetano Levanté	C48FT	2012
	FJ61EWT	Volvo B9R	Caetano Levanté	C48FT	2012
	FJ61EWV	Volvo B9R	Caetano Levanté	C48FT	2012
	FJ61EWW	Volvo B9R	Caetano Levanté	C48FT	2012

Details of the vehicles in this fleet may be found in our *English Bus Handbook : Coaches* book

Silverdale Tours is another operator to choose the Volvo B12Bs with Plaxton bodywork before the Levanté was introduced. FJ55DYW is seen at Marble Arch while working route 450. *Colin Lloyd*

The National Express Handbook

SKYLINE TRAVEL

Skyline Travel, Parsonage Street, Oldbury, B69 4PH

388	Blackpool - Burnham-on-Sea				
662	Birmingham - Skegness				
*	PO54NAA	MAN 18.360	Marcopolo Viaggio 350	C49FT	2004

No vehicles are contracted in National Express colours. The vehicles used on the services are selected from the main fleet.

SOUTH GLOUCESTERSHIRE

South Gloucestershire Bus & Coach, Pegasus Business Park, Gypsy Patch Lane, Patchway, Bristol, BS34 6QD

040	London - Bristol
200	Gatwick Airport - Bristol
318	Birmingham - Bristol
339	Grimsby - Westward Ho!
402	London - Frome
403	London - Bath
444	London - Glasgow
502	London - Ilfracombe

South Gloucestershire now undertakes many of the duties latterly operated by First from the Bristol area. Several of the initial coaches for its duties came from National Express stock with six new Scania tri-axle coaches arriving new. One of these, FJ58AKP is seen at the London end of route 040, the frequent service that links London with Bristol. *Colin Lloyd*

Another view of **FJ58AKP**, this time an off-side view of the coach as it leaves Victoria for Bristol.
Mark Bailey

FJ57KGF	Scania K340 EB6	Caetano Levanté	C61FT	2007	National Express, 2009
FJ57KGG	Scania K340 EB6	Caetano Levanté	C61FT	2007	National Express, 2009
FJ57KHH	Scania K340 EB6	Caetano Levanté	C61FT	2007	National Express, 2009
FJ58AKN	Scania K340 EB6	Caetano Levanté	C61FT	2008	
FJ58AKO	Scania K340 EB6	Caetano Levanté	C61FT	2008	
FJ58AKP	Scania K340 EB6	Caetano Levanté	C61FT	2008	
FJ58AKU	Scania K340 EB6	Caetano Levanté	C61FT	2008	
FJ58AKV	Scania K340 EB6	Caetano Levanté	C61FT	2008	
FJ58AKX	Scania K340 EB6	Caetano Levanté	C61FT	2008	
FJ60HYC	Volvo B9R	Caetano Levanté	C48FT	2010	
FJ60HYF	Volvo B9R	Caetano Levanté	C48FT	2010	
FJ60HYR	Volvo B9R	Caetano Levanté	C48FT	2010	
FJ60HYS	Volvo B9R	Caetano Levanté	C48FT	2010	
FJ60HYT	Volvo B9R	Caetano Levanté	C48FT	2010	
FJ60HYU	Volvo B9R	Caetano Levanté	C48FT	2010	
FJ60HYV	Volvo B9R	Caetano Levanté	C48FT	2010	
FJ60HYW	Volvo B9R	Caetano Levanté	C48FT	2010	
FJ60HYX	Volvo B9R	Caetano Levanté	C48FT	2010	
FJ60HYY	Volvo B9R	Caetano Levanté	C48FT	2010	
FJ60HYZ	Volvo B9R	Caetano Levanté	C48FT	2010	
FJ60HZA	Volvo B9R	Caetano Levanté	C48FT	2010	
FJ61GZE	Volvo B9R	Caetano Levanté	C48FT	2012	
FJ61GZF	Volvo B9R	Caetano Levanté	C48FT	2012	
FJ61GZG	Volvo B9R	Caetano Levanté	C48FT	2012	
FJ61GZH	Volvo B9R	Caetano Levanté	C48FT	2012	
FJ12FXF	Volvo B9R	Caetano Levanté	C48FT	2012	
FJ12FXG	Volvo B9R	Caetano Levanté	C48FT	2012	
FJ12FXH	Volvo B9R	Caetano Levanté	C48FT	2012	
FJ12FXK	Volvo B9R	Caetano Levanté	C48FT	2012	
FJ12FXL	Volvo B9R	Caetano Levanté	C48FT	2012	

Details of the vehicles in this fleet may be found in our *English Bus Handbook: Coaches* book

STAGECOACH

Stagecoach UK operates several National Express services. Management units are Cambridge (EA), Kent (SE), Oxfordshire (OX) and Yorkshire (Y).

007	London - Dover	SE
021	London - Dover	SE
022	London - Ramsgate	SE
310	Bradford - Poole	Y
312	Rotherham - Blackpool	Y
447	London - Lincoln	EA
448	London - Peterborough - Grimsby	EA
449	London - Mablethorpe	EA
465	London - Huddersfield	Y
484	London - Walton-on-the-Naze	EA
560	Barnsley - London	Y
564	London - Halifax	Y
737	Stansted Airport - Oxford	OX

Stagecoach vehicles allocated to National Express duties:

53035	EA	YN54YRJ	Volvo B12M		Plaxton Paragon Expressliner	C46FT	2004	
53037	Y	YN05XBD	Volvo B12M		Plaxton Paragon Expressliner	C46FT	2004	

53701-53716 Volvo B9R Plaxton Elite C48FT 2010

53701	OX	OU10GYH	53705	OX	OU10GYO	53709	EA	AE10JTV	53713	Y	YN60ACF
53702	OX	OU10GYJ	53706	EA	AE10JSZ	53710	EA	AE10JTX	53714	Y	YN60ACJ
53703	OX	OU10GYK	53707	EA	AE10JTO	53711	EA	AE10JTY	53715	Y	YN60ACO
53704	OX	OU10GYN	53708	EA	AE10JTU	53712	Y	YN60ABX	53716	Y	YN60ACU

59201-59215 Scania K340 EB4 Caetano Levanté C49FT 2006

59201	SE	FJ56PDX	59205	SE	FJ56OBR	59209	SE	FJ56OBN	59213	SE	FJ56OCA
59202	SE	FJ56PEO	59206	SE	FJ56OBS	59210	SE	FJ56OBX	59214	SE	FJ56OCB
59203	SE	FJ56PFA	59207	SE	FJ56OBT	59211	SE	FJ56OBY	59215	SE	FJ56OCC
59204	SE	FJ56PFB	59208	SE	FJ56OBU	59212	SE	FJ56OBZ			

59301-59305 Scania K340 EB6 Caetano Levanté C61FT 2008

59301	Y	FJ08KNY	59303	Y	FJ08KOE	59304	Y	FJ08KOH	59305	Y	FJ08KNH
59302	Y	FJ08KOA									

Details of the other vehicles in the Stagecoach fleet, along with allocation code details, may be found in the annual *Stagecoach Bus Handbook*.

Four of the Stagecoach depots provide coaches for National Express duties. Based at Peterborough are six Volvo B9Rs with Plaxton Elite bodywork, including 53706, AE10JSZ, seen here.
Mark Doggett

E STOTT & SONS

E Stott & Sons Ltd, Colne Vale Garage, Saville Street, Milnsbridge, Huddersfield, HD3 4PG

310	Bradford - Leicester
335	Halifax - Birmingham
351	Sheffield - Blackpool
542	Glasgow - Blackpool
561	Bradford - London
660	Bradford - Skegness

ES1	FJ60EGC	Volvo B9R	Caetano Levanté	C48FT	2010
ES2	FJ60EGD	Volvo B9R	Caetano Levanté	C48FT	2010
ES3	FJ12FXA	Volvo B9R	Caetano Levanté	C48FT	2012

Details of the vehicles in this fleet may be found in the *English Bus Handbook: Notable Independents* book

Based in Huddersfield, E Stott operates three coaches on National Express work. All are the current standard product including FJ60EGC shown. *Mark Bailey*

STUART'S of CARLUKE

Stuarts Coaches Ltd, Castlehill, Airdrie Road, Carluke, ML8 5EP

538	Edinburgh - Birmingham				
542	Glasgow - Blackpool				
544	Glasgow - London				
592	Glasgow - London				
	WSV238	Bova Futura FHD12-340	Bova	C49FT	2002
*	SF09HKH	VDL Bova Futura FHD127.365	VDL Bova	C53FT	2009
	FJ11MKO	Volvo B9R	Caetano Levanté	C48FT	2011
	FJ11MKP	Volvo B9R	Caetano Levanté	C48FT	2011

Previous registration:
WSV238   JW02BUS, SN02NMP

Details of the other vehicles in this fleet may be found in *The Scottish Bus Handbook*.

Stuart's of Carluke operates three coaches in full National Express livery and, like many of the operators contracted to operate coaches at weekends, provides white liveried vehicles that carry boards with the National Express name. Noted for its fleet of VDL Bova coaches, Stuart's fleet is represented by **SF09HKH**. *Colin Lloyd*

TELLINGS - GOLDEN MILLER

MK Metro - TGM Stansted

Tellings-Golden Miller (Coaches) Ltd, Electra Avenue, London Heathrow Airport, Hounslow, TW6 2DN

010	London - Cambridge	TGM	
011	London - King's Lynn	TGM	
012	London - Dereham	TGM	
302	Bristol - Northampton	MK	
337	Rugby - Brixham	MK	
455	London - Northampton	MK	
707	Gatwick Airport - Northampton	MK	

BU1-BU6   Scania K340 EB4   Caetano Levanté   C49FT   2006-07
BU1 FJ56PDO   **BU3** FJ07DWF   **BU5** FJ56PCX   **BU6** FJ56OBP
BU2 FJ56PCZ   **BU4** FJ07DWG

YN56SGO   Volvo B12B   Plaxton Panther   C49FT   2007
YN56SGU   Volvo B12B   Plaxton Panther   C49FT   2007

MK1-MK17   Scania K340 EB4   Caetano Levanté   C49FT   2006-07
MK1 FJ07TKC   **MK2** FJ07TKE   **MK3** FJ07TKF   **MK17** FJ56PDK

MK5-11   Scania K340 EB4   Caetano Levanté   C49FT   2008
MK5 MK FJ09DXG   **MK7** MK FJ09DXL   **MK9** MK FJ09DXO   **MK11** MK FJ09DXR
MK6 MK FJ09DXK   **MK8** MK FJ09DXM   **MK10** MK FJ09DXP

Details of the other vehicles in this fleet may be found in the annual *Arriva Bus Handbook*.

Part of the Arriva group, Tellings provides vehicles for the two bases now running on the National Express network. FJ09DXP is seen heading for Gatwick Airport. *Mark Doggett*

TRAVELLERS' CHOICE

Shaw Hadwin (John Shaw & Sons) Ltd, The Coach and Travel Centre, Scotland Road, Carnforth, LA5 9BQ

333	Blackpool - Bournemouth
341	Burnley - Birmingham - Paignton
570	Blackpool - London

PE56UJX	Volvo B12B	Plaxton Panther	C49FT	2006
FJ56PCV	Scania K340 EB4	Caetano Levanté	C49FT	2006
FJ56PDU	Scania K340 EB4	Caetano Levanté	C49FT	2006
FJ56PDV	Scania K340 EB4	Caetano Levanté	C49FT	2006
PO12GWG	Volvo B9R	Caetano Levanté	C48FT	2012
PO12GWJ	Volvo B9R	Caetano Levanté	C48FT	2012
PO12GWL	Volvo B9R	Caetano Levanté	C48FT	2012

Details of the other vehicles in this fleet may be found in the *English Bus Handbook : Coaches* book.

From its base near the M6 at Carnforth, John Shaw now uses seven coaches for the duties undertaken on National Express services. Recent additions are three Volvo B9R coaches represented by PO12GWJ, as it leaves London for Blackpool. *Dave Heath*

TRAVELSTAR

Travelstar European Ltd, Marlow Street, Walsall, WS2 8AQ

210	Heathrow Airport - Wolverhampton
310	Coventry - Birmingham - Leeds - Bradford
320	Birmingham - Bradford
319	Oxford - Cardiff

*	GO03ECB	Scania K114 EB4	Irizar PB	C49FT	2003	Go-Goodwin, Eccles,
*	YE52TSE	Bova Futura FHD12.340	Bova	C49FT	2004	
	FJ11GOA	Volvo B9R	Caetano Levanté	C48FT	2011	
	FJ11GOC	Volvo B9R	Caetano Levanté	C48FT	2011	
	FJ61EVY	Volvo B9R	Caetano Levanté	C48FT	2012	
	FJ61GZA	Volvo B9R	Caetano Levanté	C48FT	2012	
	FJ61GZB	Volvo B9R	Caetano Levanté	C48FT	2012	

Previous registrations:

GO03ECB	YE51TSE	YE52TSE	B1AFC, SV04GRF

Sheffield is the location for this view of Travelstar's FJ61GZB. This operator provides an early morning departure from Birmingham to the Yorkshire town on route 320 before an afternoon return. *Dave Heath*

The National Express Handbook

WHITTLES

Whittle Coach & Bus, Foley Business Park, Stourport Road, Kidderminster DY11 7QL

409	London - Aberystwyth
410	London - Birmingham - Wolverhampton
444	London - Worcester
545	London - Pwllheli

84	FJ11GJX	Volvo B9R	Caetano Levanté	C48FT	2011
85	FJ11GJY	Volvo B9R	Caetano Levanté	C48FT	2011
86	FJ11GJZ	Volvo B9R	Caetano Levanté	C48FT	2011
87	FJ11GKA	Volvo B9R	Caetano Levanté	C48FT	2011
88	FJ11GKC	Volvo B9R	Caetano Levanté	C48FT	2011
89	FJ11GLV	Volvo B9R	Caetano Levanté	C48FT	2011

Details of the other vehicles in this fleet may be found in the *English Bus Handbook : Smaller Groups* book.

Whittles, part of the EYMS Group, only recently commenced National Express operations and uses six new Volvo B9R coaches, all delivered during 2011. Representing the operation is 87, FJ11GKA, photographed shortly after its arrival. *Bill Potter*

YELLOW BUSES - RAPT

Bournemouth Transport Ltd, Yeoman's Way, Bournemouth, BH8 0BQ

032	London - Southampton
035	London - Poole - Weymouth
035	London - Bournemouth University
652	London - Southampton (Cruise Terminal)

| 319 | FJ06GGE | Volvo B12B | | Caetano Enigma | | C49FT | 2006 | |
| 320 | FJ06GGF | Volvo B12B | | Caetano Enigma | | C49FT | 2006 | |

| **321-324** | | Volvo B12B | | Caetano Levanté | | C49FT | 2007 | |
| 321 | FJ07DVY | **322** | FJ07DVZ | **323** | FJ07DWA | | **324** | FJ07DWC |

| 328 | FJ59AOV | Scania K340 EB4 | | Caetano Levanté | | C49FT | 2009 | |
| 329 | FJ59ARO | Scania K340 EB4 | | Caetano Levanté | | C49FT | 2009 | |

330-339		Volvo B9R		Caetano Levanté		C49FT	2011-12	
330	FJ60HYN	**333**	FJ61EWD	**336**	FJ61GZM		**338**	FJ12FXE
331	FJ60HYO	**334**	FJ61EWE	**337**	FJ61GZN		**339**	FJ12FXT
332	FJ61EWC	**335**	FJ61GZL					

Details of the other vehicles in this RATP fleet may be found in the *English Bus Handbook : Groups* book.

Now part of the French-owned RAPT Group, Yellow Buses operates on four National Express routes, all connecting London with the South West. Representing the fleet is 321, FJ07DVY. *Colin Lloyd*

YEOMANS

Yeomans Canyon Travel Ltd, 21-3 Three Elms Trading Estate, Hereford, HR4 9PU

222	Heathrow Airport - Hereford
444	London - Hereford
445	London - Hereford

4	FJ07TKD	Scania K340 EB4	Caetano Levanté	C49FT	2007
41	FJ08KNK	Scania K340 EB6	Caetano Levanté	C49FT	2008
42	FJ08KNX	Scania K340 EB6	Caetano Levanté	C49FT	2008
43	FJ08KNZ	Scania K340 EB6	Caetano Levanté	C49FT	2008
44	FJ08KOB	Scania K340 EB6	Caetano Levanté	C49FT	2008
45	FJ60EFZ	Volvo B9R	Caetano Levanté	C48FT	2010
46	FJ60HYG	Volvo B9R	Caetano Levanté	C48FT	2010

Details of other vehicles in this fleet may be found in the *English Bus Handbook : Notable Independents* book.

Pictured in the summer of 2012, Yeoman's tri-axle Scania K340 FJ08KNZ, is seen heading home on route 444. This coach joined the fleet in June 2008 and has since been joined by two Volvo coaches. Yeomans also requires two coaches for its 222 duties gained in 2010. *Graham Crawshaw*

54   The National Express Handbook

YOURBUS

Yourbus, Heanor Gate Industrial Estate, Heanor Gate Road, Heanor, DE75 7RJ
Classic Coaches, Morrison Road, Annfield Plain, Stanley, DH9 7RX

023	London - Bexhill
040	London - Bristol
230	Mansfield - Nottingham - Gatwick Airport
240	Bradford - Birmingham - Gatwick Airport
310	Bradford - Nottingham
324	Bradford - Paignton
425	Newcastle-upon-Tyne - London
426	South Shields - Sunderland - London
435	London - Ashington
436	London - South Shields
440	London - Derby
450	London - Nottingham
461	London - Lichfield
540	London - Colne
561	London - Bradford - Skipton
563	London - Whitby
591	London - Edinburgh
594	London - Edinburgh

4001-4016   Volvo B9R   Caetano Levanté   C48FT   2011

4001	FJ60EHN	4005	FJ60HYB	4009	FJ60EGY	4013	FJ60KVM
4002	FJ60HXW	4006	FJ60EGZ	4010	FJ60KVH	4014	FJ60KVO
4003	FJ60HXZ	4007	FJ60EFX	4011	FJ60KVK	4015	FJ60KVP
4004	FJ60HYA	4008	FJ60EFY	4012	FJ60KVL	4016	FJ60KVR

4017-4028   Volvo B9R   Caetano Levanté   C48FT   2011

4017	FJ11GLY	4020	FJ11MLZ	4023	FJ11MMF	4026	FJ11MMU
4018	FJ11GKE	4021	FJ11MMA	4024	FJ11MMK	4027	FJ11MMX
4019	FJ11MLY	4022	FJ11MME	4025	FJ11MMO	4028	FJ11MOA

Yourbus is the successor to Dunn Line and operates from two centres in Durham and Nottingham. Shown here is 4016, FJ60KVR, one of the 2011 intake of Volvo B9Rs.
Mark Doggett

4029-4043		Volvo B9R		Caetano Levanté		C48FT	2012	
4029	FJ61EXM	4033	FJ61EXR	4037	FJ61EXV		4041	FJ61EYA
4030	FJ61EXN	4034	FJ61EXS	4038	FJ61EXW		4042	FJ61EYC
4031	FJ61EXO	4035	FJ61EXT	4039	FJ61EXX		4043	FJ61EYD
4032	FJ61EXP	4036	FJ61EXU	4040	FJ61EXZ			
4101-4110		Volvo B12B		Caetano Levanté		C49FT	2007	Veolia, 2010
4101	FJ56PCF	4104	FJ56PBY	4107	FN06FLE		4109	FN06FLG
4102	FJ56PBU	4105	FJ56PBZ	4108	FN06FLF		4110	FN06FKZ
4103	FJ56PBX	4106	FN06FLB					
4201	FJ07DWL	Scania K340 EB4		Caetano Levanté		C49FT	2007	Veolia, 2010
4202	FJ07DWK	Scania K340 EB4		Caetano Levanté		C49FT	2007	Veolia, 2010
	YN54ZHK	Volvo B12B		Plaxton Panther		C49FT	2004	Burtons, 2008
	YN54ZHL	Volvo B12B		Plaxton Panther		C49FT	2004	Burtons, 2008
	YN54ZHM	Volvo B12B		Plaxton Panther		C49FT	2004	Burtons, 2008
	YN05VRU	Volvo B12B		Plaxton Panther		C49FT	2005	Burtons, 2008
	YN56SGO	Volvo B12B		Plaxton Panther		C49FT	2006	
	YN56SGU	Volvo B12B		Plaxton Panther		C49FT	2006	Tellings-GM, 2009
	FJ07DWF	Scania K340 EB4		Caetano Levanté		C49FT	2007	
	FJ07DWG	Scania K340 EB4		Caetano Levanté		C49FT	2007	
	FJ57KGK	Scania K340 EB6		Caetano Levanté		C61FT	2007	
	FJ08KMA	Scania K340 EB6		Caetano Levanté		C61FT	2008	
	FJ08KME	Scania K340 EB6		Caetano Levanté		C61FT	2008	
	FJ08KMF	Scania K340 EB6		Caetano Levanté		C61FT	2008	
	FJ08KMG	Scania K340 EB6		Caetano Levanté		C61FT	2008	
	FJ60HYH	Volvo B9R		Caetano Levanté		C48FT	2011	
	FJ60HYK	Volvo B9R		Caetano Levanté		C48FT	2011	
	FJ60HYL	Volvo B9R		Caetano Levanté		C48FT	2011	
	FJ60HYM	Volvo B9R		Caetano Levanté		C48FT	2011	

The early Caetano Levanté coaches for National Express contracts were based on Volvo B12B chassis as well as Scania products. More recently they have become available on Volvo B9R and MAN 18.360 chassis. Yourbus 4042, FJ61EYC is seen while operating route 481 to London from Lichfield. *Mark Doggett*

Index to National Express routes

007	London - Dover	333	Blackpool - Bournemouth
010	London - Cambridge	335	Halifax - Poole
011	London - King's Lynn	336	Edinburgh - Plymouth
012	London - Dereham	337	Rugby - Brixham
021	London - Dover	339	Grimsby - Westward Ho!
022	London - Ramsgate	341	Burnley - Birmingham - Southsea
023	London - Bexhill	350	Liverpool - Cambridge - Clacton
024	London - Eastbourne	351	Blackpool - Leeds
025	London - Brighton	380	Newcastle - Leeds - Liverpool
026	London - Bognor Regis	381	Leeds - Chester - Wrexham
030	London - Fareham/Southsea	384	Birmingham - Llandudno
031	London - Portsmouth	385	Manchester - Bangor
032	London - Southampton	386	Manchester - Pwllheli
033	London - Salisbury - Yeovil	387	Coventry - Blackpool
035	London - Poole/Bournemouth	388	Birmingham - Burham-on-Sea
040	London - Bristol - Burnham-on-Sea	397	Leicester - Blackpool
040	London - Bristol University	402	London - Frome
060	Leeds - Manchester- Liverpool	403	London - Bath Spa
061	Leeds - Manchester Airport	404	London - Penzance
090	London - Southend-on-Sea	406	London - Newquay
200	Gatwick Airport - Bristol	409	London - Aberystwyth
201	Gatwick Airport - Swansea	410	London - Birmingham - Wolverhampton
202	Heathrow Airport - Cardiff	417	London - Stafford
203	Heathrow Airport - Southsea	418	London - Wrexham
205	Heathrow Airport - Poole	420	London - Birmingham - Wolverhampton
206	Gatwick Airport - Poole	421	London - Blackpool
210	Gatwick - Heathrow - Wolverhampton	422	London - Burnley
210	Gatwick - Heathrow Airport - Birmingham	425	London - Newcastle - Ashington
222	Gatwick Airport - Gloucester - Hereford	426	London - Sunderland - South Shields
230	Gatwick Airport - Nottingham - Derby	435	London - Ashington
240	Gatwick Airport - Heathrow - Bradford	436	London - South Shields
250	Heathrow Airport - Ipswich	440	London - Leicester - Manchester
300	Bristol - Southsea	444	London - Gloucester
302	Bristol - Northampton	445	London - Hereford
304	Liverpool - Birmingham - Weymouth	447	London - Lincoln
305	Liverpool - Southend-on-Sea	448	London - Peterborough - Grimsby
308	Birkenhead - Great Yarmouth	449	London - Mablethorpe
310	Bradford - Nottingham - Poole	450	London - Mansfield - Mansfield
312	Chesterfield - Blackpool	455	London - Northampton
314	Southport - Birmingham - Cambridge	460	London - Strafford-upon-Avon - Coventry
315	Eastbourne - Helston	461	London - Lichfield
318	Liverpool - Birmingham - Bristol	465	London - Huddersfield
319	Bradford - Oxford	481	London - Ipswich - Felixstowe
320	Bradford - Birmingham - Cardiff	484	London - Walton-on-the-Naze
321	Bradford - Aberdare	490	London - Norwich - Great Yarmouth
322	Hull - Swansea	491	London - Lowestoft
323	Liverpool - Cardiff	497	London - Great Yarmouth
324	Bradford - Paignton	500	London - Penzance
325	Birmingham - Manchester	501	London - Totnes - Brixham
326	Newcastle - Nottingham - Cambridge	502	London - Ilfracombe
327	Scarborough - Bath Spa	504	London - Penzance
328	Rochdale - Manchester - Plymouth	507	London - Swansea
329	Nottingham - Newcastle upon Tyne	508	London - Haverfordwest
330	Nottingham - Penzance	509	London - Cardiff
332	Newcastle upon Tyne - Swindon	528	Rochdale - Birmingham - Haverfordwest

The National Express Handbook   *57*

Bruce's provides coaches for the long journeys between Aberdeen and London and from Edinburgh to Plymouth. Seen entering the National Express coach station in the city is FJ57KHV. *Mark Bailey*

530	Newcastle - Paignton	592	London - Glasgow - Aberdeen
531	Newcastle - Plymouth	594	London - Edinburgh
532	Edinburgh - Plymouth	596	London - Edinburgh
534	Glasgow - Hull	650	London - Dover (Cruise Terminal)
537	Glasgow - Corby	652	London - Southampton (Cruise Terminal)
538	Inverness - Manchester - Chester	660	Skegness (Butlins) - Bradford
538	Aberdeen - Manchester - Chester	661	Skegness (Butlins) - Coventry
538	Inverness - Coventry	662	Skegness (Butlins) - Birmingham
539	Edinburgh - Bournemouth	663	Skegness (Butlins) - Newcastle
540	London - Bolton/Bury/Burnley/Rochdale	664	Skegness (Butlins) - Liverpool
542	Glasgow - Blackpool	668	Bognor Regis (Butlins) - London
544	Glasgow - London	672	Minehead (Butlins) - Swansea
545	London - Llandudno - Pwllheli	675	Minehead (Butlins) - Wolverhampton
550	London - Birkenhead - Liverpool	701	Heathrow Airport - Woking
560	London - Sheffield - Barnsley	707	Gatwick Airport - Northampton
561	London - Bradford - Knaresborough	727	Brighton - Gatwick Airport - Norwich
562	London - Hull	737	Stansted Airport - Oxford
563	London - Whitby	747	Brighton - Gatwick - Heathrow Airport
564	London - Halifax	767	Stansted Airport - Nottingham
570	London - Blackpool	777	Stansted Airport - Birmingham
570	London - Whitehaven	787	Cambridge - Heathrow
580	Liverpool - Newcastle upon Tyne	797	Cambridge - Gatwick Airport
588	London - Inverness	A3	London - Gatwick Airport
590	London - Glasgow - Aberdeen	A6	London - Stansted Airport
591	London - Edinburgh	A9	London (Liverpool St) - Stansted Airport

Vehicle Index

2086PP	Galloway	FJ07DVP	National Express	FJ11GKC	Whittle
2513PP	Galloway	FJ07DVR	National Express	FJ11GKD	Mike DeCourcey
574CPT	Go North East	FJ07DVY	Yellow Buses	FJ11GKE	Yourbus
A17XEL	Excelsior	FJ07DVZ	Yellow Buses	FJ11GKF	Mike DeCourcey
A18XEL	Excelsior	FJ07DWA	Yellow Buses	FJ11GKG	Mike DeCourcey
A19XEL	Excelsior	FJ07DWC	Yellow Buses	FJ11GKK	Mike DeCourcey
A20XEL	Excelsior	FJ07DWF	TGM Stansted	FJ11GKL	Mike DeCourcey
AE10JSZ	Stagecoach	FJ07DWG	TGM Stansted	FJ11GKN	Mike DeCourcey
AE10JTO	Stagecoach	FJ07DWK	Yourbus	FJ11GKO	Mike DeCourcey
AE10JTU	Stagecoach	FJ07DWL	Yourbus	FJ11GKP	Rotala
AE10JTV	Stagecoach	FJ07DWN	National Express	FJ11GKU	Rotala
AE10JTX	Stagecoach	FJ07TKC	TGM Milton Keynes	FJ11GKV	Rotala
AE10JTY	Stagecoach	FJ07TKD	Yeomans	FJ11GKX	Rotala
BK10EHT	National Express	FJ07TKE	TGM Milton Keynes	FJ11GKY	Edwards Coaches
BK10EHU	National Express	FJ07TKF	TGM Milton Keynes	FJ11GKZ	Mike DeCourcey
BK10EHV	National Express	FJ08KLF	Go North East	FJ11GLF	Epsom Coaches
BK10EHW	National Express	FJ08KLS	Go North East	FJ11GLK	Rotala
BK10EHX	National Express	FJ08KLU	Go North East	FJ11GLV	Whittle
BK10EHY	National Express	FJ08KLX	Go North East	FJ11GLY	Yourbus
BK10EHZ	National Express	FJ08KLZ	Go North East	FJ11GLZ	Rotala
BK10EJU	National Express	FJ08KMU	Go North East	FJ11GME	Mike DeCourcey
BK10EJV	National Express	FJ08KMV	Go North East	FJ11GMF	Mike DeCourcey
BK10EJX	National Express	FJ08KMX	Peter Godward	FJ11GMG	Mike DeCourcey
BK10EJY	National Express	FJ08KNH	Stagecoach	FJ11GMO	Edwards Coaches
BK10EJZ	National Express	FJ08KNK	Yeomans	FJ11GMU	Edwards Coaches
BU08CGG	Excelsior	FJ08KNV	Go North East	FJ11GMV	Epsom Coaches
BU53AWN	National Express	FJ08KNW	Go North East	FJ11GMX	Edwards Coaches
BU53AWO	National Express	FJ08KNX	Yeomans	FJ11GMY	Edwards Coaches
BU53AWP	National Express	FJ08KNY	Stagecoach	FJ11GMZ	Edwards Coaches
BU53AWR	National Express	FJ08KNZ	Yeomans	FJ11GNF	Edwards Coaches
BU53AWT	National Express	FJ08KOA	Stagecoach	FJ11GNK	Edwards Coaches
BU53AWV	National Express	FJ08KOB	Yeomans	FJ11GNN	Edwards Coaches
CU6860	Go North East	FJ08KOE	Stagecoach	FJ11GNO	Edwards Coaches
FD54DGV	Ambassador	FJ08KOH	Stagecoach	FJ11GNP	Excelsior
FD54DHX	Ambassador	FJ09DWY	Silverdale	FJ11GNU	Excelsior
FD54DHY	Ambassador	FJ09DXA	Ambassador	FJ11GNV	Excelsior
FH05URN	National Express	FJ09DXB	Ambassador	FJ11GNX	Excelsior
FH05URO	National Express	FJ09DXC	Ambassador	FJ11GNY	Excelsior
FH05URP	National Express	FJ09DXE	Ambassador	FJ11GNZ	Excelsior
FH05URR	National Express	FJ09DXG	TGM Milton Keynes	FJ11GOA	Travelstar European
FH06EAW	National Express	FJ09DXK	TGM Milton Keynes	FJ11GOC	Travelstar European
FH06EAX	National Express	FJ09DXL	TGM Milton Keynes	FJ11GOH	Edwards Coaches
FH06EBL	National Express	FJ09DXM	TGM Milton Keynes	FJ11MJK	National Express
FH06EBM	National Express	FJ09DXO	TGM Milton Keynes	FJ11MJO	National Express
FH06EBN	National Express	FJ09DXP	TGM Milton Keynes	FJ11MJU	National Express
FH06EBO	Selwyns	FJ09DXR	TGM Milton Keynes	FJ11MJV	National Express
FH06URO	National Express	FJ09DXT	Silverdale	FJ11MJX	National Express
FJ05AOV	Chenery	FJ09DXU	Silverdale	FJ11MJY	National Express
FJ05AOX	Chenery	FJ10EZP	Kings Ferry	FJ11MKA	Chenery
FJ06GGE	Yellow Buses	FJ10EZR	Kings Ferry	FJ11MKC	Chenery
FJ06GGF	Yellow Buses	FJ10EZT	Silverdale	FJ11MKD	National Express
FJ06GGK	National Express	FJ10EZU	Silverdale	FJ11MKE	National Express
FJ06URG	National Express	FJ10EZV	National Express	FJ11MKF	National Express
FJ06URH	National Express	FJ11GJO	Rotala	FJ11MKG	National Express
FJ07DVH	National Express	FJ11GJU	Rotala	FJ11MKK	National Express
FJ07DVK	National Express	FJ11GJV	Mike DeCourcey	FJ11MKL	National Express
FJ07DVL	National Express	FJ11GJX	Whittle	FJ11MKM	National Express
FJ07DVM	National Express	FJ11GJY	Whittle	FJ11MKN	National Express
FJ07DVN	National Express	FJ11GJZ	Whittle	FJ11MKO	Stuarts Carluke
FJ07DVO	National Express	FJ11GKA	Whittle	FJ11MKP	Stuarts Carluke

The National Express Handbook   59

Approaching Marble Arch is FJ07TKF operating from the TGM fleet and initially new to Arrive The Shires as fleet number 4103. Route 455 plies between London and Northampton with the early evening departure extended to Corby. *Dave Heath*

FJ11MKU	National Express	FJ12FXD	National Express	FJ12FYL	National Express
FJ11MKV	National Express	FJ12FXE	Yellow Buses	FJ12FYM	National Express
FJ11MKZ	Mike DeCourcey	FJ12FXF	South Gloucestershire	FJ12FYN	National Express
FJ11MLE	Rotala	FJ12FXG	South Gloucestershire	FJ12FYO	National Express
FJ11MLF	Rotala	FJ12FXH	South Gloucestershire	FJ12FYP	National Express
FJ11MLK	Rotala	FJ12FXK	South Gloucestershire	FJ12FYR	National Express
FJ11MLL	Rotala	FJ12FXL	South Gloucestershire	FJ55DYW	Silverdale
FJ11MLN	Rotala	FJ12FXM	Edwards Coaches	FJ55DZK	Silverdale
FJ11MLO	Rotala	FJ12FXO	Edwards Coaches	FJ56OBP	TGM Stansted
FJ11MLU	Mike DeCourcey	FJ12FXP	Edwards Coaches	FJ56OBR	Stagecoach
FJ11MLV	Selwyns	FJ12FXR	Edwards Coaches	FJ56OBS	Stagecoach
FJ11MLY	Yourbus	FJ12FXS	Edwards Coaches	FJ56OBT	Stagecoach
FJ11MLZ	Yourbus	FJ12FXT	Yellow Buses	FJ56OBU	Stagecoach
FJ11MMA	Yourbus	FJ12FXU	Edwards Coaches	FJ56OBW	Stagecoach
FJ11MME	Yourbus	FJ12FXV	Edwards Coaches	FJ56OBX	Stagecoach
FJ11MMF	Yourbus	FJ12FXW	Edwards Coaches	FJ56OBY	Stagecoach
FJ11MMK	Yourbus	FJ12FXX	Edwards Coaches	FJ56OBZ	Stagecoach
FJ11MMO	Yourbus	FJ12FXY	Edwards Coaches	FJ56OCA	Stagecoach
FJ11MMU	Yourbus	FJ12FXZ	Edwards Coaches	FJ56OCB	Stagecoach
FJ11MMX	Yourbus	FJ12FYA	Edwards Coaches	FJ56OCC	Stagecoach
FJ11MOA	Yourbus	FJ12FYB	Edwards Coaches	FJ56PAO	Selwyns
FJ11RDO	National Express	FJ12FYC	Edwards Coaches	FJ56PBF	Selwyns
FJ11RDU	National Express	FJ12FYD	Edwards Coaches	FJ56PBO	Selwyns
FJ11RDV	National Express	FJ12FYE	Edwards Coaches	FJ56PBU	Yourbus
FJ11RDX	Lucketts	FJ12FYF	Edwards Coaches	FJ56PBX	Yourbus
FJ11RDY	Lucketts	FJ12FYG	Edwards Coaches	FJ56PBY	Yourbus
FJ12FGK	Edwards Coaches	FJ12FYH	National Express	FJ56PBZ	Yourbus
FJ12FXA	Stotts	FJ12FYJ	Edwards Coaches	FJ56PCF	Yourbus
FJ12FXC	Edwards Coaches	FJ12FYK	Edwards Coaches	FJ56PCV	Travellers Choice

August 2012 and National Express 114, FJ60HXX is seen at Attleborough in its Commonwealth Games white livery. The matching gold-liveried version is illustrated in the 7th edition of this Handbook.
Colin Lloyd

FJ56PCX	TGM Stansted	FJ57KHO	National Express	FJ58AKU	South Gloucestershire
FJ56PCZ	TGM Stansted	FJ57KHR	National Express	FJ58AKV	South Gloucestershire
FJ56PDK	TGM Milton Keynes	FJ57KHT	National Express	FJ58AKX	South Gloucestershire
FJ56PDO	TGM Stansted	FJ57KHU	Bruce's	FJ58AKY	Bruce's
FJ56PDU	Travellers Choice	FJ57KHV	Bruce's	FJ59AOV	Yellow Buses
FJ56PDV	Travellers Choice	FJ57KHW	Bruce's	FJ59AOZ	Lucketts
FJ56PDX	Stagecoach	FJ57KHX	National Express	FJ59APF	Lucketts
FJ56PEO	Stagecoach	FJ57KHY	National Express	FJ59APX	Selwyns
FJ56PFA	Stagecoach	FJ57KHZ	National Express	FJ59APY	Selwyns
FJ56PFD	Stagecoach	FJ57KJA	National Express	FJ59APZ	Premiere
FJ56PFE	National Express	FJ57KJE	National Express	FJ59ARF	Premiere
FJ56PFF	National Express	FJ57KJF	Bruce's	FJ59ARO	Yellow Buses
FJ57 KHP	National Express	FJ57KJO	National Express	FJ60EFP	Bennetts
FJ57KGE	Bruce's	FJ57KJU	National Express	FJ60EFR	Bennetts
FJ57KGF	South Gloucestershire	FJ58AHE	Lucketts	FJ60EFS	Bennetts
FJ57KGG	South Gloucestershire	FJ58AHF	Lucketts	FJ60EFT	Bennetts
FJ57KGY	Bruce's	FJ58AHG	Lucketts	FJ60EFU	Selwyns
FJ57KGZ	National Express	FJ58AHN	Lucketts	FJ60EFV	Selwyns
FJ57KHA	National Express	FJ58AJO	Silverdale	FJ60EFW	Selwyns
FJ57KHB	Bruce's	FJ58AJU	Selwyns	FJ60EFX	Yourbus
FJ57KHC	National Express	FJ58AJV	Selwyns	FJ60EFY	Yourbus
FJ57KHD	National Express	FJ58AJX	Lucketts	FJ60EFZ	Yeomans
FJ57KHE	National Express	FJ58AJY	Lucketts	FJ60EGC	Stotts
FJ57KHF	National Express	FJ58AKF	Lucketts	FJ60EGD	Stotts
FJ57KHG	National Express	FJ58AKG	Lucketts	FJ60EGE	Johnson Bros
FJ57KHH	South Gloucestershire	FJ58AKK	Bruce's	FJ60EGF	Johnson Bros
FJ57KHK	National Express	FJ58AKN	South Gloucestershire	FJ60EGY	Yourbus
FJ57KHL	National Express	FJ58AKO	South Gloucestershire	FJ60EGZ	Yourbus
FJ57KHM	National Express	FJ58AKP	South Gloucestershire	FJ60EHB	Excelsior

The National Express Handbook   61

FJ60EHC	Excelsior	FJ61EWV	Silverdale	HSK659	Parks of Hamilton
FJ60EHD	Excelsior	FJ61EWW	Silverdale	HSK660	Parks of Hamilton
FJ60EHE	Excelsior	FJ61EWX	Premiere	JCN822	Go North East
FJ60EHF	Excelsior	FJ61EWY	Premiere	K2VOY	Go North East
FJ60EHN	Yourbus	FJ61EWZ	Selwyns	K3VOY	Go North East
FJ60HXS	National Express	FJ61EXK	Selwyns	KSK948	Parks of Hamilton
FJ60HXT	National Express	FJ61EXL	Selwyns	KSK949	Parks of Hamilton
FJ60HXU	National Express	FJ61EXM	Yourbus	KSK950	Parks of Hamilton
FJ60HXV	National Express	FJ61EXN	Yourbus	KSK951	Parks of Hamilton
FJ60HXW	Yourbus	FJ61EXO	Yourbus	KSK952	Parks of Hamilton
FJ60HXX	National Express	FJ61EXP	Yourbus	KSK953	Parks of Hamilton
FJ60HXY	National Express	FJ61EXR	Yourbus	KSK984	Parks of Hamilton
FJ60HXZ	Yourbus	FJ61EXS	Yourbus	KSK985	Parks of Hamilton
FJ60HYA	Yourbus	FJ61EXT	Yourbus	KSK986	Parks of Hamilton
FJ60HYB	Yourbus	FJ61EXU	Yourbus	KX58GTU	National Express
FJ60HYC	South Gloucestershire	FJ61EXV	Yourbus	KX58GTY	National Express
FJ60HYF	South Gloucestershire	FJ61EXW	Yourbus	KX58GTZ	National Express
FJ60HYG	Yeomans	FJ61EXX	Yourbus	KX58GUA	National Express
FJ60HYN	Yellow Buses	FJ61EXZ	Yourbus	KX58GUC	National Express
FJ60HYO	Yellow Buses	FJ61EYA	Yourbus	KX58GUD	National Express
FJ60HYR	South Gloucestershire	FJ61EYC	Yourbus	KX58GUE	National Express
FJ60HYS	South Gloucestershire	FJ61EYD	Yourbus	KX58GUF	National Express
FJ60HYT	South Gloucestershire	FJ61EYF	Premiere	KX58GUG	National Express
FJ60HYU	South Gloucestershire	FJ61EYG	Premiere	KX58GUH	National Express
FJ60HYV	South Gloucestershire	FJ61EYH	Premiere	KX58GUJ	National Express
FJ60HYW	South Gloucestershire	FJ61EYK	Epsom Coaches	KX58GUK	National Express
FJ60HYX	South Gloucestershire	FJ61EYL	Epsom Coaches	KX58GUL	National Express
FJ60HYY	South Gloucestershire	FJ61GZA	Travelstar European	KX58GUM	National Express
FJ60HYZ	South Gloucestershire	FJ61GZB	Travelstar European	LJ12LLJ	Llew Jones
FJ60HZA	South Gloucestershire	FJ61GZE	South Gloucestershire	LK53KVX	National Express
FJ60KVH	Yourbus	FJ61GZF	South Gloucestershire	LK53KVY	National Express
FJ60KVK	Yourbus	FJ61GZG	South Gloucestershire	LK53KVZ	National Express
FJ60KVL	Yourbus	FJ61GZH	South Gloucestershire	LK53KWD	National Express
FJ60KVM	Yourbus	FJ61GZL	Yellow Buses	LK53KWE	National Express
FJ60KVO	Yourbus	FJ61GZM	Yellow Buses	LSK506	Parks of Hamilton
FJ60KVP	Yourbus	FJ61GZN	Yellow Buses	LSK507	Parks of Hamilton
FJ60KVR	Yourbus	FN06FKZ	Yourbus	LSK510	Parks of Hamilton
FJ60KVS	National Express	FN06FLB	Yourbus	LSK513	Parks of Hamilton
FJ61EVN	Galloway	FN06FLE	Yourbus	LSK611	Parks of Hamilton
FJ61EVP	Galloway	FN06FLF	Yourbus	LSK613	Parks of Hamilton
FJ61EVR	Galloway	FN06FLG	Yourbus	LSK830	Parks of Hamilton
FJ61EVT	Galloway	FN06FLH	National Express	LSK831	Parks of Hamilton
FJ61EVU	Galloway	FN06FMA	National Express	LSK832	Parks of Hamilton
FJ61EVV	Galloway	FN06FMC	National Express	LSK835	Parks of Hamilton
FJ61EVW	Lucketts	FN07BYV	National Express	LSK839	Parks of Hamilton
FJ61EVX	Go South Coast	FN07BYW	National Express	LSK845	Parks of Hamilton
FJ61EVY	Travelstar European	FN07BYX	National Express	LSK870	Parks of Hamilton
FJ61EWA	Lucketts	FN07BYZ	National Express	MX61BBF	National Express
FJ61EWB	Selwyns	FN07BZA	National Express	MX61BBJ	National Express
FJ61EWC	Yellow Buses	FN07BZB	National Express	MX61BBK	National Express
FJ61EWD	Yellow Buses	FN07BZC	National Express	MX61BBN	National Express
FJ61EWE	Yellow Buses	HSK642	Parks of Hamilton	N999RWC	Chenery
FJ61EWF	Selwyns	HSK644	Parks of Hamilton	OU10GYH	Stagecoach
FJ61EWG	Selwyns	HSK645	Parks of Hamilton	OU10GYJ	Stagecoach
FJ61EWH	Selwyns	HSK646	Parks of Hamilton	OU10GYK	Stagecoach
FJ61EWK	Selwyns	HSK651	Parks of Hamilton	OU10GYN	Stagecoach
FJ61EWL	Selwyns	HSK652	Parks of Hamilton	OU10GYO	Stagecoach
FJ61EWM	Selwyns	HSK653	Parks of Hamilton	PE56UJX	Travellers Choice
FJ61EWN	Chalfont	HSK654	Parks of Hamilton	PO12GWG	Travellers Choice
FJ61EWO	Chalfont	HSK655	Parks of Hamilton	PO12GWJ	Travellers Choice
FJ61EWP	Silverdale	HSK656	Parks of Hamilton	PO12GWL	Travellers Choice
FJ61EWR	Silverdale	HSK657	Parks of Hamilton	SN08AAU	National Express
FJ61EWT	Silverdale	HSK658	Parks of Hamilton	SN08AAV	National Express